MW01717485

MINITAB® MANUAL

Peter Zehna

The Naval Postgraduate School

Introductory Statistics

Sixth Edition

and

Elementary Statistics

Fifth Edition

NEIL A. WEISS

Addison Wesley

Boston San Francisco New York
London Toronto Sydney Tokyo Singapore Madrid
Mexico City Munich Paris Cape Town Hong Kong Montreal

Reproduced by Addison-Wesley from files supplied by the author.

Printed in the United States of America.

ISBN 0-201-73712-4

1 2 3 4 5 6 7 8 9 10 VG 04 03 02 01

PREFACE

The purpose of this manual is to guide the reader in the use of the computer in the solution of statistical problems. It is designed to be used with the computer package MINITAB®.† The manual is best used in conjunction with either of the two books, Introductory Statistics, 6/e or *Elementary Statistics 5/e* both by Neil A. Weiss (Addison-Wesley, Boston). Indeed, each of the lessons found in this manual are keyed to appropriate sections in these textbooks. Nevertheless, the manual may easily be modified for use with almost any introductory textbook dealing with the standard topics examined here.

The primary focus in this manual is on Release 13 of Minitab. For the purposes of solving problems in the aforementioned textbooks, earlier releases of Minitab designed to be used with Microsoft® Windows can be applied with very little modification. In addition, we have included all of the session commands corresponding to Windows menu choices and that makes the tools discussed here available for use by practically any release of Minitab.

One *special* aspect of using the computer for computations is that students can repeatedly process a formula—or several formulas—in a model situation. This approach allows them to see effects of changes immediately. Minitab's macro capability is especially helpful in this regard. The macro facility is versatile enough to allow you to create some of the few routines that are not already part of the standard Minitab release. We exploit this feature of Minitab in several places in the manual and thus provide the user with models for creating macros.

The author would like to thank Neil Weiss for providing excellent textbooks to write supplementary material for and also for his generosity and support in supplying the necessary TeX macros as well as appropriate files and the use of various examples found in the textbook itself. A debt of gratitude is extended to all the staff at Addison-Wesley with special mention of Rachel Reeve and Joe Vetere.

Monterey, California P.W.Z.

† Henceforth, we will refer to this software package as Minitab to avoid conflicts with other uses of capital letters in this manual.

CONTENTS

LESSON 12

INFERENCES FOR POPULATION PROPORTIONS 169

LESSON 13

CHI-SQUARE PROCEDURES 177

LESSON 14

DESCRIPTIVE METHODS IN REGRESSION AND CORRELATION 191

LESSON 15

INFERENTIAL METHODS IN REGRESSION AND CORRELATION 201

LESSON 1

Getting Started

GENERAL OBJECTIVE

In this lesson we introduce you to some preliminary material that is necessary for utilizing Minitab, Release 13 for Windows. You will learn how to access Minitab and how to use this manual for solving problems in the two books *Introductory Statistics, sixth edition* (hereinafter referred to as *IS/6e*) and *Elementary Statistics, fifth edition* (hereinafter referred to as *ES/5e*) both by Neil Weiss.† You will learn how to enter data, edit the data for corrections, and execute Minitab's command language. Additionally, we introduce you to Minitab files and Minitab's powerful macro capability.

LESSON OUTLINE

1.1 Introduction
1.2 Minitab commands
1.3 Data input/output
1.4 Minitab files
1.5 Simple random sampling
1.6 Systematic random sampling
1.7 Conclusion

† Published by Addison-Wesley, Boston

1.1 Introduction

In this supplement you will learn how a computer can be used to solve statistics problems. The most commonly used programs for statistical work are taken from *statistical software packages.* Statistical software packages are collections of statistical computer programs written by some organization or individual. Specifically, we will examine how the statistical software package MINITAB† can be applied to perform the statistical analyses discussed in the books *IS/6e* and *ES/5e*. To get the most from this supplement, you should use one of these textbooks while at the computer and do the examples and problems as you read.

For ease of reference, we will call this supplement "the manual" and either textbook, *IS/6e* or *ES/5e* , "the book or the textbook." The manual is divided into *lessons.* We have used the word "lesson" instead of "chapter" to avoid confusion with the book. The first four sections of this lesson, Lesson 1, can be read immediately. They do not require any knowledge of statistics.

Most lessons in the manual correspond to chapters in the book having the same number. That is, Lesson 2 of the manual discusses how Minitab can be applied to solve the statistical problems considered in Chapter 2 of the book; Lesson 3 discusses using Minitab to solve the statistical problems considered in Chapter 3 of the book; and so on. Exceptions occur in distinguishing between the two books *IS/6e* and *ES/5e*. We will now explain how we accommodate that distinction.

Look on page 23 of this manual. To the right of the section heading "1.5 Simple random sampling" you will see a box with numbers **E-1.3 : I-1.3** inside of it (**E** for *Elementary Statistics* and **I** for *Introductory Statistics.*). These entries indicate the section number of the two books that reference the material to be covered in the section of this manual; eventually, the two entries will be different. When the references refer to different sections in the two books covered by this manual, you will see the difference in the appropriate box. This means that you will benefit most in the manual when you have read the sections referenced for the appropriate book.

There are many versions of Minitab currently in use. In this manual, we focus on Release 13 for Windows (abbreviated R13 in this manual). We assume the reader is familiar with the basics of using Windows on the computer. This includes the use of a mouse as well as familiarity with menus, dialog boxes and the general manipulation of windows.

Other versions of the program are available on a variety of mainframes, minicomputers, and microcomputers. Hence the actual procedure for accessing Minitab files can vary. To obtain the necessary information, consult your computer manual, your instructor, or an appropriate member of the staff at your computer center.

Fortunately, all versions of Minitab have a certain core of instructions that are common. Although our main focus here is on the use of Windows and Release 13, we have added instructions called *session commands* which may be typed at the `MTB >` prompt in Minitab (see next section) in lieu of the windows-oriented menu

† As stated in the preface, we will refer to this software package as Minitab to avoid conflicts with other uses of capital letters in this manual.

commands. This makes it possible to use virtually any version of Minitab by following the appropriate typing instructions that are given in this manual. If your version of Minitab is rather recent (Releases 9–12, or the last two releases of *The Student Edition of Minitab*) both sets of commands will work, some with only a slight modification that you will be able to find in your reference manuals. But in almost all cases, the session commands, if available at all, will accomplish the task using even the earliest versions of the software.

1.2 Minitab commands

Once you have accessed Minitab, the computer will display a sign-on message that will vary somewhat according to the implementation you are using. Figure 1.1 shows a typical screen for Release 13 for Windows.†

FIGURE 1.1 Minitab sign-on screen

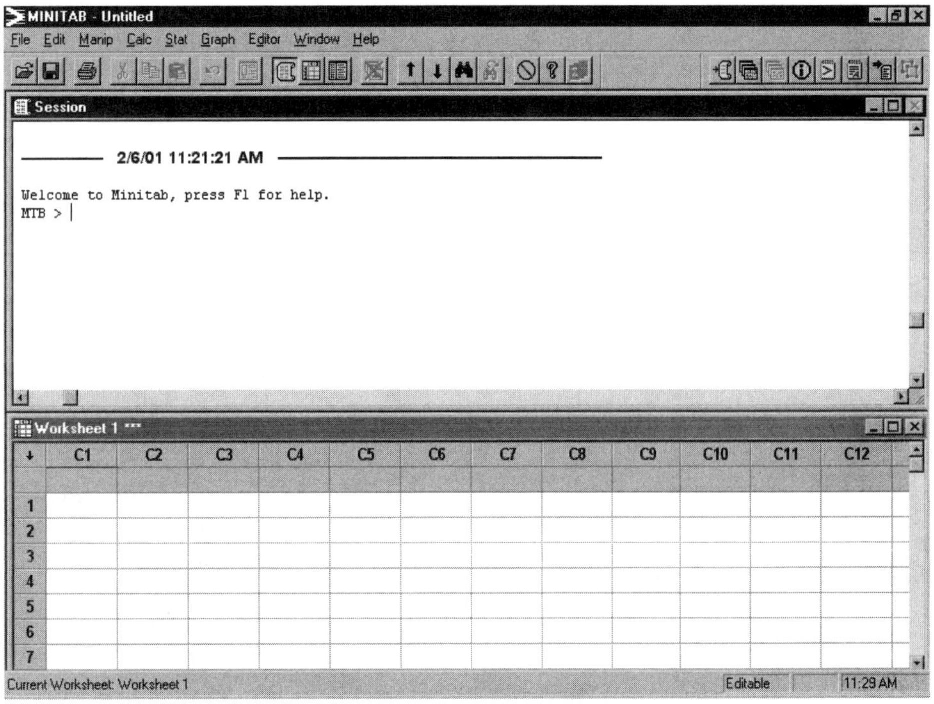

In this particular figure you see two separate windows opened for you. A window, as you see in the figure, is simply a framed region on the screen. Each window may contain, among other things, borders, a title bar, and a menu bar located just

† What you actually see will depend upon how you configured your particular installation of Minitab.

under the title bar. The top window is called the *Session* window. In it, you see the standard elements of any Windows screen starting with the *title* bar, followed by the *menu* bar, followed by the *toolbar* consisting of icons to assist in carrying out windows operations. In Figure 1.1 observe that there are nine menu choices in the menu bar. These are displayed as File , Edit , Manip , Calc , Stat , Graph , Editor , Window , and Help , with one letter of each underlined. Clicking on one of these with the left mouse button, or pressing the key combination Alt with the underlined letter simultaneously on the keyboard, will produce a drop down menu for additional options. Directions to open the file menu, for example, would appear in the manual as Alt + F.

The way we have configured our computer, the Session window appears on top of a second one called the *Data* window, but many other options exist. Not visible here is yet another window called the *Project Manager* window which shows the collection of various components of Minitab currently active. The windows that are currently available for your view may be discovered by selecting Window in the Menu bar of the Session window. Clicking any one of the choices in the dropdown menu will reveal that particular window. In this manual we only highlight those elements of Windows that impact on the manual in a particular way. For added details you should consult your Windows and Minitab references as well as the online help facilities.

SESSION WINDOW

The Session window is reserved for typing commands and viewing output. When the session window is activated, its title bar is highlighted (as in Figure 1.1). and then the Editor menu contains commands relevant to the session window. Among those is one option (the third one in the list, labeled Enable Commands) which has to do with enabling or disabling the session command language. To use this manual in the most effective manner, we suggest that you click this option so that it has a check mark next to it. You can always disable this choice later if you prefer by clicking this option again to clear the check box.

In Minitab, *commands* are instructions or computer routines for accomplishing certain tasks. You can execute these instructions *directly* by typing the appropriate command language at the `MTB >` prompt in the Session window and then pressing the Enter↵ key. We call these *session commands*. This capability is present in all releases of Minitab. An example of a session command as it will appear in this manual is the following (ignore the meaning for now):

Session commands:

- Type `Correlation C7 C8` at the `MTB >` prompt and press Enter↵

Minitab provides a device called the *Command Line Editor*, accessible under Edit in the Session window for assistance in entering commands. However, for our present purposes there is little need for exploiting that alternative and we will provide all of our session commands as illustrated above.

MENU COMMANDS

You can also execute these instructions *indirectly* by using a variety of Windows menus provided in R13. These menus organize commands by function rather than command language names. Through the use of typical Windows submenus and dialog boxes the appropriate command language is executed. Indeed, the corresponding session commands are echoed in the Session window as though you typed them (unless this feature has been disabled). We call these menu choices *menu commands* and they are the primary focus in your book. However, as previously remarked, by including session commands as well, we make virtually all Minitab procedures in this manual available to all users of Minitab, regardless of the release they may own.

If you have a mouse connected to your system, you may access, or *select*, these menus by moving the mouse pointer to the desired menu option and tapping the left mouse button. This action is referred to as *clicking on* the menu name. In all of the cases where we give menu instructions in this manual, you have the option of using the mouse to click on the resulting choice. We will not always mention this explicitly in the instructions because it will always be understood.

Each menu choice can also be selected by pressing the Alt key along with the key represented by the underlined letter. For example, you can access the excellent on-line help facility by pressing the Alt at the same time that you press the H key –denoted Alt+H in this manual (or just press F1.) Doing so will bring up a help screen with several topics called *submenus*. These submenus can be accessed by clicking on the submenu with the mouse or pressing the direction arrow key ↓ on your keyboard a sufficient number of times to highlight the submenu, then pressing Enter↵ . (Use ↑ to reverse the direction.)

To illustrate, let us select the Stat menu. Assume you are in the Session window. Just click on this choice with the mouse or else press Alt+S.

FIGURE 1.2
Stat menus

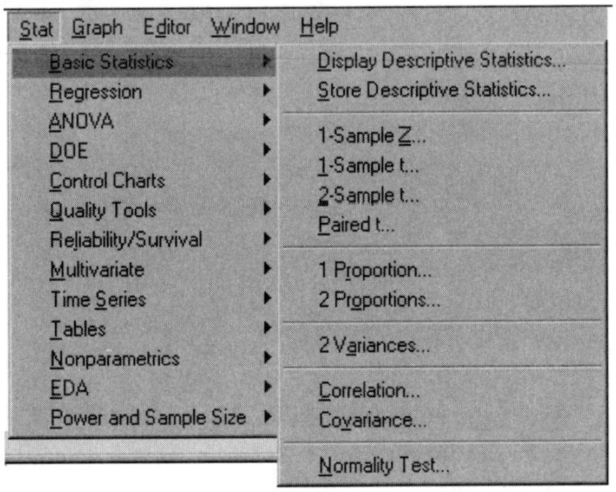

You should then see a menu like that displayed in the left panel of Figure 1.2. First, notice that the Stat button is depressed in the menu bar to remind you which menu you are in. Next, observe that there are several submenus within this menu. The symbol ► appearing at the end of a submenu indicates further submenus are available. You may *select* one of these submenus by clicking on the submenu with the mouse or simply slide the mouse pointer to that selection. You may also press the Alt key with the indicated underlined key or else press the down arrow key ↓ a sufficient number of times to highlight the submenu and then press the Enter↵ key. For instance, if you *select* the Basic Statistics submenu by any of these methods as we did you should see a display resembling that of the right panel of Figure 1.2.

There you see the submenu along with the original menu. Observe that the selection Basic Statistics is now highlighted to remind you which submenu you are in. Like the main menu, this submenu has several choices, the specifics of which you need not be concerned with at the present time. The *ellipses* dots occuring at the end of a submenu indicate that a *dialog box* follows.

Let us illustrate by selecting the Correlation... dialog box. You press Alt + C or slide the mouse pointer over into the submenu, down to the selection which will then be highlighted, then click on it with the mouse or press Enter↵ . Figure 1.3 shows a sample screen with a dialog box, waiting for input from you the user. The cursor is positioned in the Variables: text box where the names of columns of data should be entered.

FIGURE 1.3 Correlation dialog box

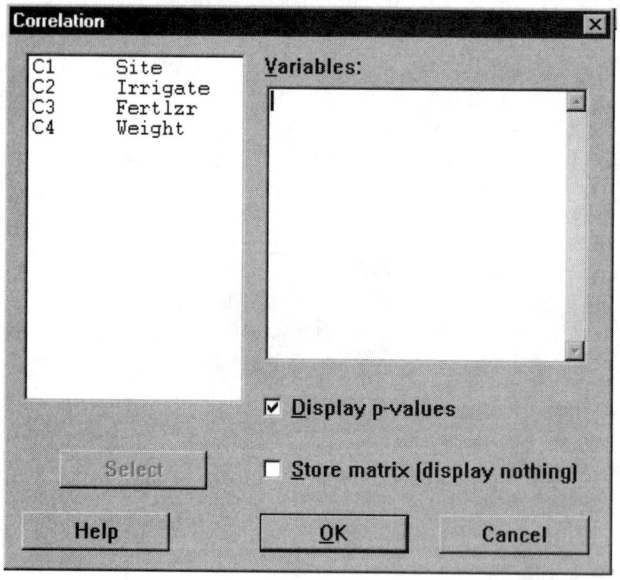

Following the convention used in R13, we will abbreviate this sequence of menu commands in this manual as follows.

Menu commands:

1 Choose Stat ➤ Basic Statistics ➤ Correlation...

This sequence means: First select the Stat menu (by whatever means) *and then* select the Basic Statistics submenu (by whatever means) *and then* select the Correlation... dialog box (by whatever means). The symbol ➤ is used as a separator and is read "*and then select*" if you like. Notice the typeface used for menu selections.

To continue with this example, the material displayed in the left-hand box, called a *list box*, will vary. This one contains the names and column identities of four different columns of data. Some dialog boxes contain other boxes to provide further information and elicit input from you, the user, in order to effect a particular command. As a general precaution, you may press the [F3] key to clear any previous entries from a dialog box. We will illustrate these other boxes as we proceed.

The mouse pointer and the [Tab] key are particularly useful in getting around different parts of a dialog box. From now on, when we say "Select such and such a location or dialog box (usually given by name)" we mean point the mouse and click or else press [Tab] a sufficient number of times to arrive at that place. Incidentally, pressing [Shift]+[Tab] reverses the order of traversing in a dialog box. You may also use the directional arrow keys as well as the mouse pointer to move among various places in the dialog box.

The cursor should be inside the Variables: box, called a *text box*, where the program is waiting for you to enter variables from the list box. You may do this in several ways. You may double click on any name. You may enter the list box by pressing [F2], use the arrow keys of the keyboard to highlight the variable of interest, then select that variable by pressing [F2] again. You may also click once on a name and click the [Select] command button located under the list box. Finally, you may type the variable name or its column designation in the Variables: text box directly.

In any particular application, we give you explicit instructions on which options to select and what input to enter, leaving the method of input to you. Notice that if you press [Tab] a number of times you find yourself at a *command button* labeled [OK] (with the letter O underlined). If you press [Tab] again, you find yourself at the [Cancel] button. Try pressing the key combination [Shift]+[Tab] to return to previous places. Incidentally, pressing [Esc], the escape key, is equivalent to clicking on [Cancel] and you will be returned to the Session window. The [Esc] key is often useful for returning to previous menus (or the Session window) from a given menu.

In this manual the expression "Select [OK]" will mean that you are to click on the [OK] button with the mouse, or else press [Tab] a sufficient number of times to arrive at the [OK] button and press [Enter], or else press [Alt]+[O]. The result is an *output*. We call the display of the Session window output a *printout*. An example can be seen in Printout 1.1.

PRINTOUT 1.1
Correlation output

```
MTB > Correlation  'Fertlzr' 'Weight'.

Correlations: Fertlzr, Weight

Pearson correlation of Fertlzr and Weight = 0.536
P-Value = 0.000
```

Whenever you select [OK], Minitab generates the command language corresponding to the options you selected by returning you to the Session window and executing the corresponding session commands. This printout is the result of selecting Fertlzr and Weight as variables in the Correlation dialog box of Figure 1.3 and selecting [OK]. (You may ignore the meaning for now.) Notice that the session commands are displayed as if you had typed them from the MTB > prompt at the outset. Indeed, sometimes, it is simpler to do so.

That much of an overview of the use of menus will suffice for now. We should point out that many menu commands have been stored in the Minitab toolbar as well. These are executed by simply clicking on the appropriate button or icon in the toolbar.† For additional details you should consult your Minitab User's Guide, as well as your reference for the use of Windows. You will be getting a lot of practice in the rest of the manual because directions are given for both menu commands and session commands in nearly every example we present.

Here are some typographical conventions that we use throughout the manual. In referring to Minitab prompts and certain other visuals *as you would see them on your screen*, we will use a typeface like the Minitab prompt MTB >. Examples are DATA> and SUBC>. Directions to you, the reader, to type a command in order to achieve a certain effect will be shown in the same typeface but underlined. Thus you might see "Type PRINT C1" (without quotes) as a direction to first type the material above the underline at the MTB > prompt *and then* press the [Enter↵] key. When we want to refer to various Minitab commands and other computer visuals *outside the context* of a display or an image of what you would see on the screen, we use the typeface PRINT. Finally, we occasionally use italic type for emphasis.

Thus, for example, we use DESCRIBE when referring broadly to this Minitab command. If we were directing you to a screen output line in which this command occurred, we would refer to its occurrence in that display as DESCRIBE. On the other hand, if we are directing you to type this command, you will see the instruction: Type DESCRIBE and generally, it will be understood that you follow this by pressing the [Enter↵] key, even when we fail to mention it explicitly.

Commands in Minitab are not case-sensitive. Thus DESCRIBE, Describe, and describe all achieve exactly the same results. Incidentally, only the first four letters of a command are used by Minitab although it does no harm to include more. For instance, whether you type DESC 'TODAY' or DESCRIBE 'TODAY', you will achieve exactly the same result. In earlier releases, Minitab allowed you to enter superfluous language in the command statement, such as, the more lengthy

DESCRIBE the data named 'TODAY'

as a session command. This is no longer the case and users of releases prior to R13 should be aware of that in order to avoid error messages. Fortunately, such messages indicate that it is probably the superfluous language that is causing the error. In this manual, essential commands and other visuals are denoted in capital letters, and at least the first four letters of commands must be typed, although you need not

† Placing the mouse pointer over the various buttons will reveal the name associated with each one.

type more than that. Whenever you have any doubt about the exact requirements of a command structure, use the Help facility in Minitab or else consult the User manuals that you received for R13.

1.3 Data input/output

In this section, we describe how to enter data into the Minitab program and how to view the data on the screen. We begin with a procedure for entering data.

THE DATA WINDOW

The later releases of Minitab, including R13, have an excellent data editor, called the *Data window*, for entering and manipulating data. In Figure 1.1, it appears as a (non-active) window below the Session window and is currently empty. To access the Data window, click on it with the mouse or press Ctrl+D from any other window, or else click on the Current Data Window button in the toolbar. Notice that the Session title bar fades and the Data title bar is highlighted to remind you of the window currently active and its current name. Press Ctrl+M when you want to return to the Session window, click anywhere in the window with the mouse, or click on the Session Window button in the toolbar.

In Figure 1.4 we display a typical Data window with some contents. If you are familiar with a spreadsheet, you will be comfortable with the Data window. Indeed, Minitab refers to this workspace as a *Worksheet*, in this case named Cereal.MTW. However, this is not a true spreadsheet in the sense that you cannot enter formulas to perform operations; you may only store data.

FIGURE 1.4
Data window

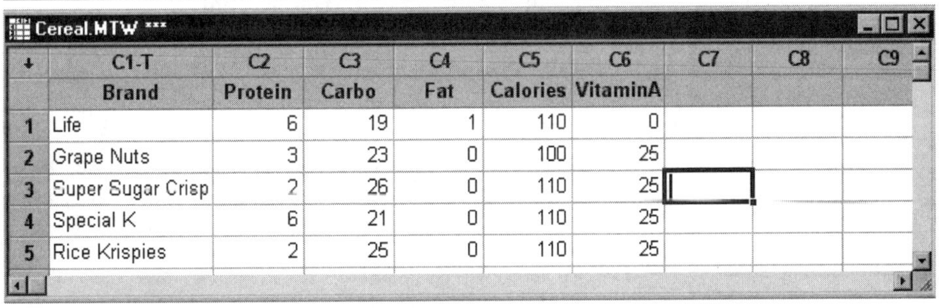
Cereal.MTW ***

↓	C1-T	C2	C3	C4	C5	C6	C7	C8	C9
	Brand	Protein	Carbo	Fat	Calories	VitaminA			
1	Life	6	19	1	110	0			
2	Grape Nuts	3	23	0	100	25			
3	Super Sugar Crisp	2	26	0	110	25			
4	Special K	6	21	0	110	25			
5	Rice Krispies	2	25	0	110	25			

To begin, you need to understand that, with Minitab, a data set is stored in a *column*. A column is designated by a C followed by a number. For example, C1 stands for column 1, C2 stands for column 2, and so forth; rows are numbered. Although the emphasis in Minitab is on columns, we will find that some row operations are recognized as well. A *cell* is defined by the intersection of a column and a row. You make a cell active by clicking on it with a mouse or by using the direction arrow keys. The active cell is highlighted with a dark rectangular border around it (row 3, column 7 in Figure 1.4) and has a thick (blinking) cursor, waiting for you to enter data. Notice that column 1 is designated C1-T indicating that the entries are text rather than numeric.

Notice also that the second row in the worksheet (the one between the column and row designations) has been reserved for naming the column containing data. Thus you see the name Brand (capital letters are not important) for C1-T. A name may be up to 31 characters long (only up to 8 characters in some older releases). Any characters with the exception of a single quote, the pound symbol,or the asterisk, and a beginning or ending blank space may be used. You may also name any column in the Session window by using the session command NAME. Press Ctrl+M to return to the Session window. Typing `Name C2 'Protein'` at the MTB > prompt would attach the name to the column. Note the use of the single quotes around the name. Any time a name is referred to in a Session window, it must be enclosed in single quotes. You cannot use the same name for two variables in the same worksheet.

Many of the data sets to be used with this manual have been recorded in the Minitab folder of the *WeissStats CD* released to you with your textbook and given a name according to the context in which the data are used in *IS/6e* and *ES/5e*. In addition, unnamed raw data are supplied in a text format in a folder named Text Files. Occasionally, in Lessons 3–5 in particular, you will be asked to enter data not found on the CD. Hence, you should become acquainted with some data entry fundamentals for the entry of data.

If you must enter data manually, the Data window is by far the easiest method. To enter the numerical data shown in C2, for example, you have only to access the cell defined by row 1 and column C2 by pointing the mouse pointer to that location. (The mouse pointer changes shape to an open cross ✣). Make sure that the direction arrow in the upper left cell of the worksheet is pointing downwards; if not click it to change direction. Then type 6. Press ↓ or Enter↵ . Then type 3, and proceed in that manner to the last entry.

THE SET COMMAND

Out of deference to those readers who may not have some of the later versions of Minitab, we review some session commands that can be used to input data. The SET command is used to input one data set at a time, and the READ command is used to input two or more data sets of the same size simultaneously.

EXAMPLE 1.1 Illustrates the SET session command

Table 1.1 on the following page gives the number of days to maturity for 40 short-term investments. The data are from *Barron's National Business and Financial Weekly*. Input this data set into a Minitab worksheet by using the computer.

TABLE 1.1 Days to maturity for 40 short-term investments

70	64	99	55	64	89	87	65
62	38	67	70	60	69	78	39
75	56	71	51	99	68	95	86
57	53	47	50	55	81	80	98
51	36	63	66	85	79	83	70

SOLUTION For those readers who have access to the *WeissStats CD* you may find the same data in a worksheet file named Tb02-01. In the menu commands choose

File ➢ Open Worksheet ➢ Data Sets ➢ Minitab Files ➢ Problems ➢ Chapter_02 ➢ Tb02-01

In the future, we will merely mention that this or that worksheet file is located in the *WeissStats CD* by writing this as (WeissStats CD:Tb02-01) and you may follow this sequence of commands to enter it into your Minitab workspace for use if you wish.

To enter the same data manually, open a new worksheet by selecting File ➢ New and then choose Minitab Worksheet as an option. To store the data in C2 in the Session window, type SET C2 at the MTB > prompt and press [Enter↵]. The prompt MTB > is displayed on the screen to indicate that data may now be entered. Now type in the numbers in the data set, separating the numbers by commas or spaces (any number of spaces will do as long as the numbers are separated). When you have finished entering the data, type END and then press [Enter↵].

If you make a mistake entering data, you can immediately type END, retype SET C2, and start all over again. The computer will erase your old data. Generally speaking, whenever the SET command is used, the computer deletes all numbers previously stored in the column under consideration. Name this column of numbers DAYS using the NAME command. The procedure is shown in Printout 1.2.

PRINTOUT 1.2 The SET command

```
MTB > SET C2
DATA> 70 64 99 55 64 89 87 65
DATA> 62 38 67 70 60 69 78 39
DATA> 75 56 71 51 99 68 95 86
DATA> 57 53 47 50 55 81 80 98
DATA> 51 36 63 66 85 79 83 70
DATA> END
MTB > NAME C2 'DAYS'
```

The computer now has the days-to-maturity data stored in C2 under the name DAYS. those data will be available for you to work with until you type SET C2 again or sign off the computer. ■

You can also store a single numeric value in a storage area called a *constant* using the LET command. A constant is designated by a K followed by a number. For example, K2 stands for a constant numbered 2. Let us store the number 10 in the constant K4. Proceed as follows.

Menu commands:

1 Choose Calc ➤ Calculator...
2 Type K4 in the Store result in variable: text box
3 Select the Expression: text box and type 10
4 Select OK

You will see the session command Let K4=10 echoed in the Session window which might be a simpler to accomplish the same thing. Either way, the number 10 is stored in the constant K4 in the worksheet.

You may name this constant only with a session command called NAME. For example, typing Name K4 'ten' would attach the name ten to that constant. Incidentally, you can erase the name of a column or constant by renaming the variable or using two single quotes together (not a double quote). For example, typing Name K4 '' would remove the name from K4.

PROBLEM 1.1 A research physician conducted a study on the ages of persons with diabetes and obtained the following data for the ages of a sample of 35 diabetics.

48	41	57	83	41	55	59
61	38	48	79	75	77	7
54	23	47	56	79	68	61
64	45	53	82	68	38	70
10	60	83	76	21	65	47

Enter these data into an empty column by using the Data window or the session command SET. Name the data set DIABETES.† ◇

THE READ COMMAND

To input two or more data sets of the same size simultaneously, you may of course enter the data directly into the Data window or use the session command SET twice. The session command READ is an alternative and is more efficient than the SET command for more than one column of data. The next example shows how you implement this command.

EXAMPLE 1.2 Illustrates the READ session command

Table 1.2 on the following page displays data on age (in years) and price (in hundreds of dollars) for a sample of cars. We will refer to the car as the Orion automobile, even though the data were gathered for a real car (from the *Asian Import* edition of the *Auto Trader* magazine). The data are also located in the *WeissStats CD* as a worksheet file named Tb14-02.

† The solutions to all problems can be found in Appendix A. Incidentally, problems end with the symbol ◇; the symbol ■ is used to signal the end of an example.

Input both the age data and the price data into the computer simultaneously using the READ command. Use the names AGE and PRICE for the appropriate columns.

TABLE 1.2
Age - price data for Orions

Car	Age (yrs) x	Price (\$100s) y
1	5	85
2	4	103
3	6	70
4	5	82
5	5	89
6	5	98
7	6	66
8	6	95
9	2	169
10	7	70
11	7	48

SOLUTION To store the age and price data simultaneously, use the following session command.

Session commands:

- Type READ C4 C5 at the MTB > prompt and press [Enter↵]

Then proceed as indicated in Printout 1.3.

PRINTOUT 1.3
The READ command

```
MTB > READ C4 C5
DATA> 5 85
DATA> 4 103
DATA> 6 70
DATA> 5 82
DATA> 5 89
DATA> 5 98
DATA> 6 66
DATA> 6 95
DATA> 2 169
DATA> 7 70
DATA> 7 48
DATA> END
   11 ROWS READ
MTB > Name C4 'AGE' C5 'PRICE'
```

As with SET, the READ command causes the DATA> prompt to appear. You then enter the data row by row. That is, you type 5 85, press [Enter↵], type 4 103, press [Enter↵], and so on. It is important that you press [Enter↵] after each *pair* of numbers since two columns were placed in the command. After entering the last pair of numbers, type END. Finally, type Name C4 'AGE' C5 'PRICE'. The age data from the second column of Table 1.1 are now stored in C4 and the price data from the third column of Table 1.1 are now stored in C5 under those names. ■

PROBLEM 1.2 The continental statistics in the following table provide approximate land areas and 1988 population estimates. [SOURCE: Population Reference Bureau, Inc.] Enter the land area and population data in two empty columns using the READ command or the Data window.

Continent	Land area (1000 sq mi)	Population (millions)
Asia	10,644	2,995
Africa	11,707	623
North America	9,360	416
South America	6,883	286
Europe	1,905	497
Oceana	3,284	26

Note: Minitab uses a comma to separate numbers, so you should not type the commas. For example, Minitab would view the entry 10,644 as a pair of numbers: 10 and 644. The entry 10,644 2,995 would be viewed as four numbers: 10, 644, 2, and 995. Minitab will inform you that you have entered too many values and will display the error message: `* ERROR * TOO MANY VALUES - REENTER ROW.` ◇

DISPLAYING WORKSHEETS

To see the results of your commands, as well as the current status of your worksheet, you switch windows by choosing Window ➤ Project Manager and look in the folders named Columns and Constants. Alternatively, simply type INFO as a session command. You will see a display of all of the columns and constants currently in your worksheet, along with any names.

You can see the actual contents of columns and constants in your worksheet by following the steps illustrated in the next example.

EXAMPLE 1.3 *Illustrates displaying data*

Display both the age data and the price data simultaneously for the Orion cars, given in Table 1.2 on page 13.

SOLUTION To display the data, proceed as follows.

Menu commands:

1 Choose Manip ➤ Display Data...

2 Select the variables C4, C5 from the list box by any of the methods previously discussed

3 Select [OK]

The results are displayed in Printout 1.4.

PRINTOUT 1.4
Displaying data

Row	AGE	PRICE
1	5	85
2	4	103
3	6	70
4	5	82
5	5	89
6	5	98
7	6	66
8	6	95
9	2	169
10	7	70
11	7	48

Alternatively, you may use the session command PRINT in the Session window. This command does not actually *print* on a printer for you; it merely echoes or displays the results on your screen. It takes the form PRINT E...E, where E may be replaced by any column or constant designation. In this case, execute this simple session command.

Session commands:

- Type `PRINT C4 C5` at the `MTB >` prompt and press [Enter↵]

The result will be the same. ■

PROBLEM 1.3

In Problem 1.1 on page 12, you stored some age data in a column named DIABETES. Have Minitab display the data. ◇

LEAVING MINITAB

If you exit Minitab without any safeguards, all your work will be lost, although you will be prompted to be certain that this action is intentional on your part. We discuss various ways of saving your work in a later section. In the meantime, if you want to save your work so that you may reuse it later, just type `SAVE` at the `MTB >` prompt and then press [Enter↵]. Your current worksheet will be saved in a special file which is given the name MINITAB with the file type .MTW, as shown in Printout 1.5 on the following page. When you do this, you will see that the name of the worksheet in its titlebar, if visible, changes to read MINITAB.MTW.

PRINTOUT 1.5 Saving and retrieving the MINITAB.MTW worksheet

```
MTB > save
Saving file as: C:\Program Files\MTBWIN\Data\MINITAB.MTW
* NOTE  * Existing file replaced.

MTB > retrieve
Retrieving worksheet from file: MINITAB.MTW
# Worksheet was saved on Fri Feb 09 2001
MTB >
```

Later, when you sign on again, you need only type RETRIEVE at the MTB > prompt and your worksheet will reappear just as you left it. See Printout 1.5 again. You may then continue. You may periodically type SAVE as a precaution if you like. Each time, the current status of your work is preserved in the file MINITAB.MTW, overwriting the current file as indicated in the * NOTE * in Printout 1.5.

You may now leave Minitab by choosing File ➤ Exit as a menu command or else just type Stop as a session command. Either way, a dialog box giving you a save option will be opened. Select No for now (we will discuss this save option later). Incidentally, you may notice a Save Project option in the File menu. This option, together with the save button 🖫 in the toolbar, will save all worksheets together with the session window contents and any graphs in a special folder called Project Manager. In any given session, you may always access this window by selecting it from the Window selection in the menu bar or by pressing Ctrl+I. This is a very convenient way to classify and save all of your output in one place. In this manual we focus entirely on worksheets so we have no use for this special housekeeping option and we refer the reader to the Minitab references for further details.

If you are going to do all of Lesson 1 in a single session, it would be advisable to keep the data sets from the examples in the worksheet, as we refer to them in additional illustrations. For now, you can preserve data by using the SAVE and RETRIEVE session commands that we just discussed. Remember that the special file MINITAB.MTW is overwritten each time you type SAVE as a session command.

1.4 Minitab files

Before we discuss the macro capability in Minitab, let us summarize some of the details about files in Minitab. As you will see, a macro may be viewed as a special kind of file.

Minitab has several different types of files that you may use for permanent storage of results. We use the generic term FNAME to designate the name of a file. Like columns, FNAME may be any name of your choice, usually from one to eight characters (up to 31 characters in R13). However, Minitab does not like to see files using the same name as that given a column. So in general, you should select unique, though perhaps related, names to distinguish files from columns. If you save files on a disk drive different from the drive where the Minitab program is located, you will have to add the name of the disk drive in what follows. For example, you will have to use A:FNAME or perhaps C:\ABC\FNAME if your files are located in drive A: or in the folder ABC of drive C: .

WORKSHEET FILES

First, let us discuss the one file that is not an ASCII, or ordinary text, file and that is the worksheet file. We have already discussed the SAVE command without a name specification on page 15. This session command originally creates and then continues to overwrite the default file MINITAB.MTW. Once a file has a name, you may also choose File ➤ Save Current Worksheet... in the Session window and the worksheet will be saved under that name, overwriting the original contents. If the worksheet is untitled or you wish to save the worksheet under a different name, you may execute the following menu commands. You may even change the extension or filetype but we do not consider such alternatives here.

Menu commands:

1 Choose File ➤ Save Current Worksheet As...
2 If necessary, select a folder in the Save in: text box (you may wish to use the drop down list)
3 Type FNAME in the File name: text box
4 Select the [Save] button

This saves the file with an extension of .MTW in a folder on your hard disk. You may wish to use a different file type in the Save as type: text box but .MTW is the default type which you will normally accept. All of the worksheets in the Minitab Files of the *WeissStats CD* have this file type.

To accomplish the same result with a session command, proceed as follows.

Session commands:

- Type SAVE 'FNAME' at the MTB > prompt and press [Enter↵]

Either way, a snapshot is taken of the current worksheet. The data in the columns (and their names) and constants in use at that time are saved in a file whose name is FNAME. This can be any name of your choosing. Again, avoid names that are already in use by columns.

If you do not use a name after the SAVE session command, then Minitab will choose MINITAB for the name of the file. It will also use that same name the next time you neglect to provide a name and this will erase the first file that you saved. It is a good idea to select a name of your own. As we remarked earlier, this file is not an ordinary text file so you cannot read it or edit it outside the Minitab worksheet. Its primary purpose is to have a record of your data for future use.

EXAMPLE 1.4 *Illustrates the SAVE session command*

Let us return to the maturity data in Table 1.1 on page 11. The data were stored in C2 and named DAYS in Section 1.3. Save these data in a Minitab file named INVESTS.

SOLUTION In Printout 1.6, we demonstrate typing the SAVE command to save these data in a file called INVESTS, accepting the Minitab default extension of .MTW.

PRINTOUT 1.6
The SAVE command

```
MTB > SAVE'INVESTS'
Saving file as: C:\Program Files\MTBWIN\Data\INVESTS.MTW
```

The file is now safely stored as INVESTS.MTW. ■

Any time later that you want to *replace* the current worksheet with the one that you saved as FNAME.MTW, proceed as follows.

Menu commands:

1 Choose File ➤ Open Worksheet...
2 Type FNAME in the File name: text box to retrieve the file FNAME.MTW
3 Select the [Open] button

In the Session window, you may achieve the same result with the following session command.

Session commands:

- Type `RETRIEVE 'FNAME'` at the `MTB >` prompt and press [Enter↵]

Please observe once again that the current worksheet is lost when you retrieve another one.

EXAMPLE 1.5 *Illustrates the RETRIEVE session command*

Let us return to the maturity data in Example 1.4. To make our point, we first clear the worksheet by typing `RESTART` at the `MTB >` prompt and verify that it is empty. To verify that the worksheet is indeed empty, choose Window ➤ Project Manager and select worksheets or else type `Info` at the `MTB >` prompt in the Session window as shown in Printout 1.7.

PRINTOUT 1.7
The RESTART and INFO commands

```
MTB > RESTART
MTB > INFO
 The worksheet is empty.
MTB >
```

You see that the worksheet has been cleared by the message in Printout 1.7.

Now, with all columns empty, use the menu choices we described earlier to bring back the worksheet that was saved or else type `RETRIEVE 'INVESTS'` in the Session window as shown in Printout 1.8.

PRINTOUT 1.8
The RETRIEVE and PRINT commands

```
 MTB > RETRIEVE 'INVESTS'
Retrieving worksheet from file: INVESTS.MTW
# Worksheet was saved on Sat Feb 10 2001
MTB > PRINT C1

 DAYS
    70   64   99   55   64   89   87   65   62   38   67   70   60
    69   78   39   75   56   71   51   99   68   95   86   57   53
    47   50   55   81   80   98   51   36   63   66   85   79   83
    70
```

To verify that the maturity data are in C1, we used the PRINT command. Keep in mind that the RETRIEVE command always replaces the current worksheet so you must plan accordingly. ■

PROBLEM 1.4 Repeat the examples of this section for the data in Problem 1.1 on page 12. Recall that DIABETICS is the name of the worksheet file and verify that the column name DIABETES is preserved when you retrieve the file. ◇

DATA FILES

All other files in Minitab are ASCII or text files. You may read and edit them with an ordinary text editor. If you are using a microcomputer word processor, you need to use the text editing feature of that word processor. (See the manual for your particular word processor for details on how to do so.) Indeed, you may even create these files with such editors and bring them into the Minitab worksheet with the appropriate commands. Another advantage of text files is that you may easily transport them to and from other computer programs.

Perhaps the simplest of these text files is a *data file.* This is a file having nothing but numbers in it, arranged according to columns. These files have a default file extension of .DAT. You create such a file *within* Minitab as follows. Suppose you have data in columns C15 through C20 and you would like to preserve the data in a file named FNAME, where FNAME is a name of your choice. The file will be saved in the current folder, although you have the option of changing its location to another folder and drive.

Menu commands:

1 Choose File ➤ Other Files ➤ Export Special Text...
2 Enter C15-C20 in the Columns to export: text box
3 Now select [OK] to open the Export Data To File dialog box
4 Type FNAME (or FNAME.XXX if you want to overwrite the .DAT file type) in the File name: text box
5 Select the [Save] button

You will see the Write command echoed in the Session window as though you had typed it there. Indeed, you may find it simpler to use the following session command to begin with.

Session commands:

- Type WRITE 'FNAME' C15-C20 at the MTB > prompt and press [Enter↵]

The difference between the file FNAME.DAT and the file FNAME.MTW is that you can read the data file with a simple editor like Notepad and edit the contents. In addition, you can retrieve the data from FNAME.DAT without erasing the current worksheet. As we have said you can even create this file in the first place with your text editor. Just place the numbers in columns with one or more blanks between the columns and save it as a file with file extension .DAT.

One purpose of a data file is to be able to use it in a later session. When you want to bring the data in the columns, perhaps in a different set of columns, you proceed as follows.

Menu commands:

1 Choose File > Other Files > Import Special Text...
2 Type C25-C30 in the Store data in column(s): text box
3 Select [OK] to open the Import Text From File dialog box
4 Enter FNAME in the File name: text box
5 Select the [Open] button

You will be returned to the Session window where you will see the Read command echoed on the screen along with the directory location of the file. All of this will enter the data previously stored in C15-C20 into C25-C30 of the worksheet.
The simpler session command is

Session commands:

- Type READ 'FNAME' C25-C30

Notice that you may use different columns if you wish. Another difference over the worksheet file is that the READ command does not disturb the current worksheet as does the RETRIEVE command (provided there is no conflict in the columns used of course). Both .MTW and .DAT files will preserve your *column* data. However, names are not preserved using the WRITE command, while they are left intact using the SAVE command.

EXAMPLE 1.6 Illustrates the use of the WRITE and READ session commands

Let us return to the Orion data in Table 1.2 on page 13. Assume that these data have been entered into the worksheet in columns C2 and C3 as in Example 1.2. We will name these columns AGE and PRICE, respectively. Then we use the WRITE command (or menu instructions) to store the data into a file named ORIONS. The session commands are displayed in Printout 1.9.

PRINTOUT 1.9
The WRITE command

```
 MTB > NAME C2 'AGE' C3 'PRICE'
 MTB > WRITE 'ORIONS'C2 C3
Writing data to file: ORIONS.DAT
```

There will now be a file ORIONS.DAT in the current directory of your storage disk, provided you did not override the .DAT extension.

Next, we clear the worksheet by typing RESTART as a session command. Then we use the menu commands given on page 19 or the corresponding READ session command to bring the data into C23 and C24. We used the simpler session command as illustrated in Printout 1.10.

PRINTOUT 1.10
The READ command

```
 MTB > RESTART

 MTB > READ 'ORIONS' C23 C24
      11 ROWS READ
```

As you may verify, the data are preserved and are now in C23 and C24. However, the names were not saved with the data, as you may also verify by using the PRINT command. ■

The *WeissStats CD* included with your textbook does not contain data files but, as previously mentioned, contains worksheet files making the importing of data files somewhat redundant. Even so, each worksheet file has a companion text file located in the Text Files folder of your *WeissStats CD* . These files may be read with a simple text editor like Notepad, edited and saved for future use. Alternatively, you may use the standard Windows copy and paste technique to transfer them to your current worksheet space. The same technique may be used for text files outside the context of the manual and textbook altogether.

PROBLEM 1.5 Repeat Example 1.6 for the data of Problem 1.2 on page 14. Use the names AREA and POP for the two columns. Use CONT for the name of the data file. Verify that names are not preserved when the file is read back into the worksheet. ◇

MINITAB MACROS

Macros are yet another example of ordinary text files. The word macro comes from the Greek *makro* meaning long. We also use it as shorthand for the modern technology word *macroinstruction*, which Webster defines as a single computer instruction that stands for a sequence of operations. In its simplest form, a macro just saves you the trouble (and possible mistakes) of having to type a set of instructions over and over.

Minitab provides an extensive macro capability and permits the user to supplant the command structure in the program with additional capabilities, tailored to suit individual needs. Creating macros is not one of the goals of this manual.

Nevertheless, we do present you with a few simple ones in the material to follow. In addition, some more complicated ones have been issued to you on the *WeissStats CD* that accompanied your textbook in a folder named Minitab_Macros. The interested reader is advised to consult Chapters 28–32 in the Minitab User's Guide (R13), along with the online help available in the program itself.

While some of the older (earlier releases) forms of macros, now called *execs*, are still executable, Minitab suggests that they may not support them in the future. Currently, there are two additional forms of macros, called *global* and the more advanced *local*, that are in vogue. We refer to these more commonly as *% macros* because each requires the % symbol to invoke them. They also end in the extension .MAC and can be recognized in program folders that way.

The simplest way to create a %macro is to use your text editor to open or start a file FNAME.MAC. A local macro always begins with the heading MACRO and ends with the line ENDMACRO. In between is the name of the macro in the second line followed by the body. An example of a simple local macro is shown in Printout 1.11.

PRINTOUT 1.11
A simple local macro

```
MACRO
MUSIG c p
MCONSTANT mu sigma
MCOLUMN c p
 LET mu=SUM(c*p)
 LET sigma=SQRT(SUM(c**2*p)-mu**2)
 PRINT  MU SIGMA
ENDMACRO
```

You see in the second line, that to invoke this macro you need only type %MUSIG E E at the MTB > prompt, where the each occurrence of E can be any column in the worksheet (or their names enclosed in single quotes). We will be using this simple macro in Lesson 5 but you are not to worry about the meaning or execution at this point. Nevertheless, use your favorite text editor to copy this macro line by line and save it in the folder from which you invoke your macros as MUSIG.MAC. Notice that variables c and p are created internally to the macro (hence the name local) so that they may be replaced by any two columns invoked by the macro.

The older "exec" macros are a little more restricted in their application. They may also be created with a text editor outside of Minitab. Just type in the sequence of commands, line by line, as though they were being executed. When you have entered the last statement, type END and close the file as you normally do with your text editor. Just remember to give it file extension .MTB so that you can identify these exec files easily and distinguish them from other types. An example of a simple Exec macro that will add all the values in a particular column and print them out is shown in Printout 1.12.

PRINTOUT 1.12
A simple exec macro

```
LET K40=SUM(C40)
PRINT K40
END
```

Here, C40 is the column that must hold the data and K40 is the constant in which we will store the result. As with columns, you can view the contents of any constant, K, by typing PRINT K, hence the second line of the macro. Save this macro as the file SIMPLE.MTB. To invoke this macro any time you have data in C40, simply type EXECUTE 'SIMPLE' at the MTB > prompt and press [Enter↵]. In contrast to the local macro, this macro cannot substitute other columns or constants for C40 and K40. To execute this macro, the data must be in these two specific locations in the current worksheet.

The sequence of commands in a macro are carried out as if you were typing them in the Session window. By the way, this assumes that your macros are located in the Macros folder of the MTBWIN folder in the Program Files folder where Minitab was installed. If not, then you must replace SIMPLE with a drive designation such as, D:\FOLDER1\FOLDER2\SIMPLE, for example. Also note the single quotes around SIMPLE. These are essential for the proper execution of the macro using session commands.

Alternatively, you may use the menu commands to run an "exec" macro.

Menu commands:

1. Choose File ➤ Other Files ➤ Run an Exec...
2. Type 1 in the Number of times to execute: text box (if necessary)
3. Select the [Select File] button
4. In the Run an Exec dialog box enter SIMPLE in the File name: text box
5. Select [OK]

You may find it simpler to type the instruction in the Session window to begin with. More complicated macros are available from Minitab Inc., through the *Minitab Users Group (MUG)*.† It is a good idea to copy all macros used in this manual to the Macros folder on the hard disk where Minitab is installed. We will make that assumption in giving the directions for executing macros in this manual.

1.5 Simple random sampling

E-1.3:I-1.3

The statistical inference methods studied in an introductory statistics course are designed for use with simple random samples. Given a finite population, a *simple random sampling procedure* is one for which each possible sample of a given size is equally likely to be the one selected. A sample from a population obtained by simple random sampling is often called a *simple random sample.*

Actually, there are two types of simple random samples: namely, those obtained *with replacement* and those obtained *without replacement.* We can use Minitab to obtain a simple random sample from a population of either type. The appropriate command for sampling without replacement is called SAMPLE. To sample *with replacement*, we use the RANDOM command and this will be illustrated in later

† Contact Minitab Inc., 3081 Enterprise Drive, State College, PA 16801 for further information.

lessons when we discuss simulation. We describe in detail the use of the SAMPLE command in Minitab. This command in Minitab completely replaces the use of random-number tables as discussed in Example 1.6 of *IS/6e* and *ES/5e*. We illustrate this in the context of that example.

EXAMPLE 1.7 *Illustrates the SAMPLE command*

Recall in the example that Professor Hassett wanted to interview a random sample of 15 students from a class of 728 college-algebra students. Since the students were numbered from 1 to 728 on the registration list, he obtained the random sample by selecting 15 numbers from a table of random numbers and matched the numbers to the students to be interviewed. Use Minitab to accomplish the same task.

SOLUTION First, store the population of 728 "names" in a column named POPN. The easiest way to enter the integers 1 through 728 into this column is to use *patterned* data entry. Minitab allows you to enter this sequence using only the first and last entry. Proceed as follows.

Menu commands:

1 Choose Calc ➢ Make Patterned Data ➢ Simple Set of Numbers...
2 Enter POPN in the Store patterned data in: text box
3 Select the From first value: text box and type 1
4 Select the To last value: text box and type 728
5 Select [OK]

Session commands:

- Name a column POPN
- Type `SET 'POPN'` and press [Enter]
- Type `1:728` at the `DATA>` prompt and press [Enter]
- Type `END` at the next `DATA>` prompt and press [Enter]

Using either method, the successive integers from 1 through 728 are now stored in POPN as you may verify by examining the Data window or by using the PRINT command. Name an empty column SMPL. Then create and store the random sample in the column named SMPL as follows.

Menu commands:

1 Choose Calc ➢ Random Data ➢ Sample From Columns...
2 Type 15 in the small Sample [ ] rows from column(s): text box
3 Press [Tab] and enter POPN in the large text box
4 Select the Store samples in: text box and enter SMPL
5 Select [OK]

Session commands:

- Type SAMPLE 15 'POPN' 'SMPL' at the MTB > prompt and press Enter

Now the registration numbers for 15 students will be stored in SMPL. You may view the result by examining the Data window or using the PRINT command. See Printout 1.13 for our result, which will probably differ from yours.

PRINTOUT 1.13 Minitab output of SAMPLE

```
SMPL
297   336   565  727   223  566  526  440 670  392  206
644   563   121   94
```

The students whose registration numbers correspond to those in this list would then have been interviewed. ■

PROBLEM 1.6 Refer to Example 1.7. Use Minitab to obtain your own random sample of 15 students to be interviewed. ◇

1.6 Systematic random sampling

E-1.4:I-1.4

Example 1.7 of *IS/6e* and *ES/5e* illustrates the procedure for constructing a systematic random sample from the population of 728 students treated in the last example. The only time that the random number generator is used in that technique is to select the first number (from 1 through 48) to be included in the sample. All of the other numbers are obtained by adding the same increment to the previous selection.

On the other hand, if you were called on to use this technique many times, you might be interested in automating the procedure somewhat by creating a macro. For instance, in Example 1.7 of *IS/6e* and *ES/5e*, the number 22 was the starting point. For another application, you would find the starting value using the techniques of Example 1.7 in this manual but with a sample size of 1 instead of 15. Then the first row of C4 (named SMPL in the example) will hold this value, which we may identify as C4(1) or 'SMPL'(1) in Minitab. After that, each row, k+1, of SMPL will differ from the value in the previous row, k, by the same constant.

We assume that columns named POPN and SMPL exist in the active worksheet. *If they do not, then you must create them in order for this macro to function.* For the sake of generality, we will use K30 to store an arbitrary row of the column SMPL and K31 to store the constant difference between rows (48 in our example).

Using any text editor (like Microsoft Notepad), first create an "exec" macro we call FILLER by typing in the lines shown in Printout 1.14 on the following page and saving it as FILLER.MTB in the folder where you store Minitab macros on your computer.

PRINTOUT 1.14
A macro for filling rows of a column

```
LET 'SMPL'(K30+1)='SMPL'(K30)+K33
LET K30=K30+1
END
```

A single execution of this macro adds the constant in K33 to the value currently located in row number K30 and places it in row number K30+1 which is the next row. Then the row number is increased by 1 for the next execution. To use this filler macro, we would need to start with row number 1 for K30 and be sure that the initial random value (the number 22 in the example) is in this row. Then we would have to execute this macro 14 times altogether. But we can use another macro to accomplish this task as well as take care of the necessary initial housekeeping. While we are at it, we may as well create a macro that will handle any problem, not just the example we are discussing.

We call this global macro SYSTS (for systematic sample) and display the list of commands for you in Printout 1.15 to type in your text editor before saving it as SYSTS.MAC in the folder where you store Minitab macros. You must type the lines *exactly* as you see them.

PRINTOUT 1.15
A macro to create a systematic random sample

```
GMACRO
SYSTS
#
NOTE ENTER THE POPULATION SIZE, SAMPLE SIZE:
SET C50;
  FILE 'TERMINAL';
  NOBS 2.
COPY C50 K31 K32
LET K33=ROUND(K31/K32-.5)
SET 'POPN'
1:K33
END
SET 'SMPL'
END
SAMPLE 1 'POPN' 'SMPL'
LET K30=1
LET K32=K32-1
EXEC 'FILLER' K32
PRINT 'SMPL'
ENDMACRO
```

Let us examine this macro in some detail. The first SET command causes the macro to stop and seeks input from the user, namely the population size and sample size for you to enter from the keyboard. Thus you may apply this macro to any systematic sampling problem. The COPY line simply copies the population size and sample size that you enter into constants K31 and K32, respectively.

The next line that determines the value of K33 needs some explanation. Recall (Section Example 1.7 of *IS/6e* and *ES/5e*) that the initial value is determined by dividing the population size by the sample size and then rounding down. The ROUND function in Minitab always rounds up, so we can round down by reducing the number we wish to round by 0.5. Since we want to draw a random sample from this revised population, we SET the numbers 1 through the value of K33 into the column that is named POPN and finish that phase with the END statement. The next SET command applied to the column SMPL clears it without erasing the name and concludes with its own END statement.

After that, we apply the SAMPLE command to find the initial value (22 in the example in your textbook) and store it as Row 1 of column SMPL. Following that step, we initialize the constant K30 to 1 so that when the exec macro FILLER runs, it will begin by filling in Row 2 of SMPL. How many times do we have to do this? One less than the sample size so you see that the sample size, stored in K32, is reduced by 1 just before calling for the execution of the macro FILLER and that is how many times that macro is executed before the resulting contents of SMPL are printed out for you to view.

EXAMPLE 1.8 *Illustrates the use of the macro SYSTS*

To see an immediate application, take the same circumstances as those of Example 1.7 in *IS/6e* and *ES/5e*. Printout 1.16 shows you the execution and results. As you see, this is similar to the sample displayed in Table 1.7 in *IS/6e* and *ES/5e*. It happens that the initial value in our case is 46 instead of 22; otherwise, the systematic sample is constructed just as required. The initial value was selected at random, and the difference between successive values in the sample is exactly 48.

PRINTOUT 1.16
Executing the macro SYSTS in Minitab

```
MTB > %SYSTS
Executing from file: C:\PROGRAM FILES\MTBWIN\MACROS\SYSTS.MAC
ENTER THE POPULATION SIZE, SAMPLE SIZE:
DATA> 728 15
 No new data recorded.
Executing from file: C:\PROGRAM FILES\MTBWIN\MACROS\FILLER.MTB

SMPL
      46      94     142     190     238     286     334     382
     430     478     526     574     622     670     718
```

It should be noted that some items in the population may never be selected in a systematic sample. For example, the 8 students numbered 721 through 728 can never appear in a systematic sample in this example since the largest initial value possible is 48 and $14 * 48 = 672$; hence, the largest sample value possible is 720. ■

PROBLEM 1.7

Refer to Example 1.8. Use the macro SYSTS to obtain your own systematic random sample of 15 students to be interviewed. ◇

1.7 Conclusion

This ends our brief tour of the basic tools of Minitab. There are many more facets of the ones we have discussed and there are many other tools that exist, but we have no particular use for them in this manual. For additional information, you should examine a Minitab reference manual.

Throughout the rest of this manual, with few exceptions, we will present both the menu choices and the session commands that you may type for implementing the commands for each example. In a very few cases, notably in connection with macros and the LET command, we have already pointed out that it is sometimes simpler to type the session commands at the MTB > prompt in the Session window than to cycle through the various menu choices. That is especially true when the menu choices drop you back to the Session window where you essentially have to type the same material anyway. In those few circumstances in the manual, we will not offer menu commands for you to follow.

In this lesson we have introduced you to several conventions that we intend to follow throughout the manual. For ease of reference, we summarize them here at the close of this introductory lesson.

First (see page 8) we review the convention regarding instructions to type a command, such as "Type SET C1" (without quotes) in a Session window. Generally, this means that you will follow the typing by pressing the [Enter↵] key even if we fail to mention it explicitly.

Since almost all our problems begin with entering data into a column C and probably naming the column, let us agree with the following convention. By the phrase "Store or Enter the data in column C and name the column CNAME " (without quotes) we mean the following. Either copy the appropriate worksheet file from your *WeissStats CD*, or else use the Data window or the appropriate SET or READ command followed with the NAME command at the command line (at the MTB > prompt in the Session window).

Finally, let us review the matter of selecting a variable (column, named or unnamed) to enter into a Variables: text box (see page 7). With the cursor in that text box, you may access the list box by pressing [F2], use the arrow keys of the keyboard to highlight the variable of interest, then select that variable by pressing [F2] again. Alternatively, you may double click on the name with the mouse or else click once with the mouse and click the [Select] command button, or else just type C or CNAME. We use the phrase "Enter CNAME " or "Enter C " (without the quotes) to mean use any of these alternatives

LESSON 2

Organizing Data

GENERAL OBJECTIVE

In Lesson 1 you learned some basic commands that are necessary for using Minitab. You also learned how to generate samples to form data sets. Now you will see how Minitab can be employed to organize data into tables, and summarize data with graphical displays.

LESSON OUTLINE

2.1 Grouping data
2.2 Graphs and charts
2.3 Stem-and-leaf diagrams

2.1 Grouping data

E-2.2:I-2.2

We often gain insight into a set of data using various grouping methods. By grouping the data into classes, we may tally the values in each class and often calculate relative frequencies to obtain a frequency distribution. We call this a *grouped-data table*. Minitab does not have a separate command for a grouped-data table. However, we can apply several of its commands to construct such a table. This group of commands can then be collected into a macro for repeated use on stored data sets. You should find such a macro in your MTBWIN macros folder if you copied the files from the *WeissStats CD* . It should be listed under the name Group.mac, a local macro that may be used to group any stored column of data with user specifications. We show how this macro groups data in the next example.

EXAMPLE 2.1 *Illustrates grouped-data tables*

The U.S. National Center for Health Statistics publishes data on weights and heights by age and sex in *Vital and Health Statistics.* The weights in Table 2.1, given to the nearest tenth of a pound, were obtained from a sample of 18–24-year-old males. Construct a grouped-data table for these weights. Use a class width of 20 and a first cutpoint of 120.

TABLE 2.1
Weights of 37 males, age 18-24 years

129.2	185.3	218.1	182.5	142.8	155.2	170.0	151.3
187.5	145.6	167.3	161.0	178.7	165.0	172.5	191.1
150.7	187.0	173.7	178.2	161.7	170.1	165.8	
214.6	136.7	278.8	175.6	188.7	132.1	158.5	
146.4	209.1	175.4	182.0	173.6	149.9	158.6	

SOLUTION First, name any empty column WEIGHT and enter the data of Table 2.1 (WeissStats CD:Tb02-05) into that column in the usual manner. The group macro is a local macro with a single variable for input. Since it is local, we may submit any column of data to it either by column designation or by name. In the course of running, the macro will prompt you for further inputs required to construct a table. Proceed as follows.

Session commands:

- Type `%group 'WEIGHT'` and press Enter
- At the first `DATA>` prompt, type 3 and press Enter
- At the second `DATA>` prompt, type `120 20` for the selection of first cutpoint and class width, then press Enter

The result of executing the group macro is shown in Printout 2.1 on the following page. Compare this printout with Table 2.6 in *IS/6e* and *ES/5e*, a grouped-data table found by hand. We offer no menu commands for executing this macro. ■

PRINTOUT 2.1 Minitab output for the group macro

```
Grouped-data table for WEIGHT        N=  37

 Row  LowerCut  UpperCut   Freq  RelFreq  Midpoint

   1       120       140      3    0.081       130
   2       140       160      9    0.243       150
   3       160       180     14    0.378       170
   4       180       200      7    0.189       190
   5       200       220      3    0.081       210
   6       220       240      0    0.000       230
   7       240       260      0    0.000       250
   8       260       280      1    0.027       270
```

PROBLEM 2.1 The table for the maturity data of Example 1.1 is repeated here. (You should still have the data saved on your disk as well.) Copy the data (WeissStats CD:Tb02-01) into a column named DAYS if you have not done so already.

70	64	99	55	64	89	87	65
62	38	67	70	60	69	78	39
75	56	71	51	99	68	95	86
57	53	47	50	55	81	80	98
51	36	63	66	85	79	83	70

Construct a grouped-data table, using the `group` macro. Use 30 < 40 for the first class. ◇

SINGLE-VALUE GROUPING

For discrete data it is often preferable to use each single value as a "class." In that case grouping per se is not really necessary as we show in the next example.

EXAMPLE 2.2 *Illustrates single-value grouping*

In Example 2.7 of *IS/6e* and *ES/5e*, data on the number of TV sets per household for 50 randomly selected households are presented and repeated in Table 2.2 below.

TABLE 2.2 Number of TV sets in each of 50 randomly selected households

1	1	1	2	6	3	3	4	2	4
3	2	1	5	2	1	3	6	2	2
3	1	1	4	3	2	2	2	2	3
0	3	1	2	1	2	3	1	1	1
3	2	1	2	1	1	3	1	5	1

Use Minitab's TALLY command to obtain a grouped-data table for these data using classes based on a single value.

SOLUTION Enter the data into a column named TVs (WeissStats CD:Tb02-08). Then proceed as follows.

Menu commands:

1 Choose Stat ➤ Tables ➤ Tally...
2 Specify TVs in the Variables: text box
3 Select the Counts and Percents check boxes
4 Select [OK]

Session commands:

- Type `TALLY 'TVs';` at the `MTB >` prompt and press [Enter↵]
- Type `count;` at the `SUBC>` prompt and press [Enter↵]
- Type `percent.` at the `SUBC>` prompt and press [Enter↵]

The results are displayed in Printout 2.2.

PRINTOUT 2.2
Minitab distribution for number of TV sets

```
TVs  Count  Percent
  0      1     2.00
  1     16    32.00
  2     14    28.00
  3     12    24.00
  4      3     6.00
  5      2     4.00
  6      2     4.00
 N=     50
```

The output is consistent with the fact that each numerical value (under `TVs`) constitutes a class. Compare with Table 2.9 in *IS/6e* and *ES/5e*. ■

PROBLEM 2.2 The U. S. Bureau of the Census conducts nationwide surveys on characteristics of American households. Below are data on the number of persons per household for a sample of 40 households.

2	5	2	1	1	2	3	4
1	4	4	2	1	4	3	3
7	1	2	2	3	4	2	2
6	5	2	5	1	3	2	5
2	1	3	3	2	2	3	3

Construct a grouped-data table based on single-value grouping using Minitab's TALLY command. Use a column named PERSONS for the data. ◇

2.2 Graphs and charts

E-2.3:I-2.3

Minitab has commands that will simultaneously group a data set and display the result in graph called a *frequency histogram*. We will now illustrate using the days-to-maturity data that we considered previously (Problem 2.1). In addition to grouping the data for us, this command will produce a graph for a visual aspect of the data.

EXAMPLE 2.3 Illustrates the HISTOGRAM command

The number of days to maturity for 40 short-term investments, given in Problem 2.1, is repeated here as Table 2.3.

TABLE 2.3
Days to maturity for 40 short-term investments

70	64	99	55	64	89	87	65
62	38	67	70	60	69	78	39
75	56	71	51	99	68	95	86
57	53	47	50	55	81	80	98
51	36	63	66	85	79	83	70

Use Minitab to construct a frequency distribution and frequency histogram for these data based on the classes 30 < 40, 40 < 50, . . . , 90 < 100.

Menu commands:

1 Choose Graph ➤ Histogram... to arrive at the Graph variables: text box.
2 Enter DAYS in the X textbox for Graph 1 and select the Options... button
3 Select the CutPoint option from the Type of Intervals field
4 Select the Midpoint/cutpoint positions option from the Definition of Intervals field
5 Press [Tab] and type 30:100/10 in the input box (This instructs Minitab to use cutpoints from 30 to 100 and proceed in increments of 10.)
6 Select [OK] twice

The result may be seen in Figure 2.1 on the following page. The same figure will result from executing the following session commands.

Session commands:

- Type `HISTOGRAM 'DAYS';` and press [Enter]
- Type the range of values `Cutpoint 30:100/10.` at the SUBC> prompt and press [Enter]

Notice that each rectangle in the histogram of Figure 2.1 covers an interval or class, is centered over the selected midpoint, and rises to a height equal to the frequency in the interval.† In this way we obtain a pictorial representation of this

† If the options or subcommands are not used, Minitab will select its own classes and use midpoints under the histogram bars to identify the classes.

FIGURE 2.1 Histogram of DAYS data

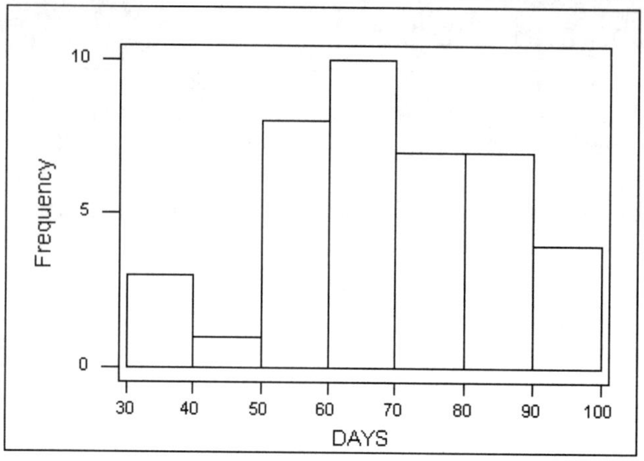

particular grouping of the data. We get a feeling of the way the data are "spread out" from low to high with a centering around 65.

This is an example of a *professional* or *high-resolution* graph. Minitab provides a very extensive graphics capability for producing a full range of quality graphs with color options and editing capabilities. For additional details beyond those we consider here, consult your Minitab for Windows references.

Minitab also provides the option of producing the simple character-style graphs found in earlier releases. The output displayed in Printout 2.3 results from the commands following the printout.

PRINTOUT 2.3 Minitab character graph histogram for DAYS data

```
Histogram of DAYS   N = 40

Midpoint   Count
    35.0       3  ***
    45.0       1  *
    55.0       8  ********
    65.0      10  **********
    75.0       7  *******
    85.0       7  *******
    95.0       4  ****
```

Menu commands:

1 Choose Graph ➤ Character Graphs ➤ Histogram...

2 Enter DAYS in the Variables: text box

3 Select the First midpoint: text box and type 35

4 Select the Interval width: text box and type 10

5 Select [OK]

In Release 13, professional or high-resolution graphics are the default; they are automatically created. To produce character graphs with *session* commands, you will have to turn off high-resolution graphs by first typing GSTD at the MTB > prompt (and turn it on again later by typing GPRO at the MTB > prompt). By contrast, the choice of Character Graphs using *Menu* commands will automatically turn on GSTD and will even turn on GPRO automatically at the conclusion, as you probably noticed. Here are the session commands that will produce the same character graph.

Session commands:

- Type GSTD at the MTB > prompt and press Enter
- Type HISTOGRAM 'DAYS'; at the SUBC> prompt and press Enter
- Type START 35; at the SUBC> prompt and press Enter
- Type INCREMENT 10. at the SUBC> prompt and press Enter

This will display the same character graph as displayed in Printout 2.3. Be sure to type GPRO at the MTB > prompt and press Enter at the conclusion of the display.

The first column of the output in Printout 2.3, headed **Midpoint**, gives the midpoint corresponding to the classes $30 < 39$, $40 < 49$, ..., $90 < 99$. The second column, headed **Count**, shows the frequencies of those classes, displayed horizontally by asterisks. However, in terms of information about how the number of days to maturity is distributed over the scale of values, the same basic information found in Figure 2.1 is conveyed by the character graph. ■

Character graphs have some advantages. They print directly in the session window and are in text format so that they are easily transported as ASCII files to other reports and applications. They are simpler to produce and often convey essentially the same information about data. Indeed, some graphs, as we shall see, are only available as character graphs in Minitab.

PROBLEM 2.3

The U.S. Bureau of Labor Statistics collects and publishes data on the ages of persons, 16 years old and over, in the civilian labor force. Suppose that the ages of a sample of 50 such persons are as follows:

22	58	40	42	43	32	34	45	38	19
33	16	49	29	30	43	37	19	21	62
60	41	28	35	37	51	37	65	57	26
27	31	33	24	34	28	39	43	26	38
42	40	31	34	38	35	29	33	32	33

Use the HISTOGRAM command and the appropriate subcommands to obtain a frequency distribution and frequency histogram of the data. Take the classes to be $15 < 20$,$20 < 25$, $25 < 30$, and so on. Name the data AGES. ◇

HISTOGRAMS FOR SINGLE-VALUE GROUPING

If we want our frequency distribution and histogram to be based on single-value classes, then we simply use a class width of 1 in setting the interval parameters.

EXAMPLE 2.4 *Illustrates a histogram for single-value grouping*

Return to the circumstances of Example 2.2 dealing with the number television sets per household.

SOLUTION We assume the data of Table 2.2 are still located in a column named TVs. If not, you will have to repeat that part of the example. To obtain a professional graph, proceed as follows.

Menu commands:

1 Choose Graph ➤ Histogram... (Press [F3] if necessary to remove any previous entries.)
2 Enter TVs in the X text box for Graph 1
3 Select the Options... button
4 Select the MidPoint radio button under Type of Intervals options
5 Select the Midpoint/cutpoint positions option button from the Definition of Intervals field
6 Press [Tab] and type 0:6/1 in the input box (This instructs Minitab to use 0 as the first midpoint and proceed by 1's to 6(to be sure and include all the data)
7 Select [OK] twice

The resulting graph is displayed in Figure 2.2.

FIGURE 2.2 Histogram of the number of TV sets per household

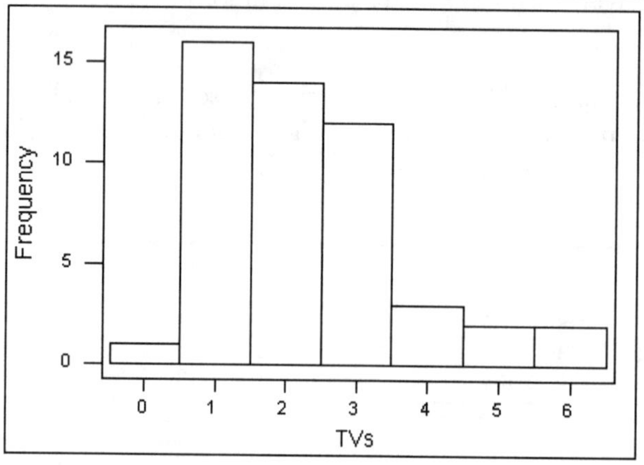

The following session commands will produce the same graph.

Session commands:

- Type HISTOGRAM 'TVs'; at the MTB > prompt and press Enter
- Type MIDPOINT. at the SUBC> prompt and press Enter

The character graph histogram may be obtained by using the subcommands START 0 and INCREMENT 1 in the session commands. If you prefer menu commands, choose Graph ➤ Character Graphs ➤ Histogram... and select TVs for the Variables: text box. See Printout 2.4 for the result.

PRINTOUT 2.4
Minitab character graph histogram for number of TV sets per household

```
Histogram of TVs   N = 50

Midpoint       Count
    0.00           1  *
    1.00          16  ****************
    2.00          14  **************
    3.00          12  ************
    4.00           3  ***
    5.00           2  **
    6.00           2  **
```

Of course, midpoints coincide in this simple case with the single values themselves since we used a class width of 1. ■

PROBLEM 2.4

The U.S. Department of Housing and Urban Development (HUD) compiles data on the number of rooms in occupied housing units. Suppose that a sample of 25 housing units yields the following data on number of rooms.

4	4	6	7	7
8	4	4	8	5
2	9	3	6	7
5	3	5	8	5
6	4	6	5	5

Employ Minitab to construct a frequency distribution and frequency histogram of the data using classes based on a single value. Name the data ROOMS. ◇

DOTPLOTS

In Example 2.12 of *IS/6e* and *ES/5e*, dotplots are discussed. Recall that a *dotplot* for a data set is constructed by displaying the possible numerical values on a horizontal axis and by recording each piece of data with a dot over the appropriate value on the horizontal axis. The Minitab command for obtaining dotplots is called DOTPLOT. This command is available for a high resolution graph in Release 13 through a local macro provided with that release. For such a simple plot, however, absolutely nothing is added to the output provided by the character graphs available in all releases including Release 13. We will obtain a dotplot for the data on oat yields considered in the next example to illustrate.

EXAMPLE 2.5 *Illustrates the DOTPLOT command*

A farmer is interested in a new fertilizer that supposedly will increase his yield of oats. He uses the fertilizer on a sample of 15 one-acre plots. The yields, in bushels, are depicted in Table 2.4.

TABLE 2.4
Oat yields

67	65	55	57	58
61	61	61	64	62
62	60	62	60	67

Use Minitab to construct a dotplot of the data.

SOLUTION First, enter the data into a column named OATS in the usual manner.

Menu commands:

1 Choose Graph > Character Graphs > Dotplot...
2 Enter OATS in the Variables: text box
3 Select OK

Session commands:

- Type DOTPLOT 'OATS' at the MTB > prompt and press Enter

See Printout 2.5 for the resulting output for both the session and menu commands.

PRINTOUT 2.5
Minitab output for DOTPLOT

```
                           .   .
   .       .   .       :   :   :       .   .       :
---+---------+---------+---------+---------+---------+---OATS
 55.00     57.50     60.00     62.50     65.00     67.50
```

If you have R13 (or R12), you should compare the dotplot in Printout 2.5 to the one you obtain from executing the high resolution option. ■

We should point out that the DOTPLOT command has two optional subcommands that allow you to control the nature of the plot. One of them, allows you to specify the numbering of the first and (optionally) the last tick mark (+) on the horizontal axis. The other allows you to specify the distance between the tick marks.

Minitab places 9 extra dashes between each tick mark so that the width between the centers of each dash represents the increment number divided by 10. In

Printout 2.5, each dash stands for 0.25 units since the increment selected by Minitab is 2.5 units on the scale. To illustrate, the plot in Printout 2.5 can be achieved as follows.

Menu commands:

1 Choose Graph ➤ Character Graphs ➤ Dotplot...
2 Enter OATS in the Variables: text box
3 Select the First midpoint: text box and type 55
4 Select the Tick increment: text box and type 2.5
5 Select OK

Session commands:

- Type DOTPLOT 'OATS'; at the MTB > prompt and press Enter
- Type START 55; at the SUBC> prompt and press Enter
- Type INCREMENT 2.5. at the SUBC> prompt and press Enter

In general, you may type START K,K in the session commands (or type the respective values of K in the First midpoint: and Last midpoint: text boxes in the menu commands) to start at the first choice of K and end at the second choice of K.

PROBLEM 2.5 Atlas Fishing Line, Inc., manufactures a 10-lb test line. A sample of 12 spools are subjected to tensile-strength tests. The results are in the following table.

9.8	10.2	9.8	9.4
9.7	9.7	10.1	10.1
9.8	9.6	9.1	9.7

Use Minitab to construct a dotplot of the data. Specify the distance between tick marks to be 1 and the first tick mark to be at 9. Name the data STRENGTH. ◇

2.3 Stem-and-leaf diagrams

E-2.4:I-2.4

Guidelines for construction and interpretation of stem-and-leaf diagrams are presented in *IS/6e* and *ES/5e*. Minitab has a command called STEM-AND-LEAF that will construct a stem-and-leaf diagram of a data set as a character graph. The command is listed directly in Graph menu as well as under Character Graphs as a submenu. It may also be found as a submenu under Stat ➤ EDA. The outcome of any choice, however, is a simple character graph. Example 2.6 indicates how to use this command.

EXAMPLE 2.6 *Illustrates the STEM-AND-LEAF command*

A pediatrician began testing cholesterol levels of young patients and was alarmed to find that a large number had levels over 200 milligrams per 100 milliliters. The readings of 20 patients with high levels are presented in Table 2.5.

TABLE 2.5
Cholesterol levels for 20 high-level patients

210	209	212	208
217	207	210	203
208	210	210	199
215	221	213	218
202	218	200	214

Use Minitab to obtain a stem-and-leaf diagram for these data using one line per stem and compare with a diagram that uses two lines per stem.

SOLUTION First, enter the data into a column named LEVELS (WeissStats CD:Tb02-18) in the usual manner. To obtain a stem-and-leaf diagram for the data using one line per stem, proceed as follows.

Menu commands:

1. Choose Graph ➤ Stem-and-Leaf...
2. Enter LEVELS in the Variables: text box
3. Select the Increment: text box and type 10
4. Select [OK]

Session commands:

- Type `STEM 'LEVELS';` at the MTB > prompt and press [Enter↵]
- Type `INCREMENT=10.` at the SUBC> prompt and press [Enter↵]

The results appear in Printout 2.6. On the first line of the output we find a verbal description of what follows (`Stem-and-leaf of LEVELS`) and the number of pieces of data (`N = 20`). The second line of the output (`Leaf Unit = 1.0`) indicates where the decimal point goes—in this case, directly after each leaf digit.

PRINTOUT 2.6
Minitab output for STEM-AND-LEAF using one line per stem

```
Stem-and-leaf of LEVELS  N  = 20
Leaf Unit = 1.0

   1   19 9
   8   20 0237889
 (11)  21 00002345788
   1   22 1
```

Let us now discuss in detail the stem-and-leaf diagram in Printout 2.6. The second column of numbers gives the stems, and to the right of the stems are the leaves. Since the leaves on each line are ordered, we see that Printout 2.6 actually provides us with an *ordered* stem-and-leaf diagram.

The numbers in the first column of the stem-and-leaf diagram in Printout 2.6, called *depths,* are used to display cumulative frequencies. Starting from the top, the depths indicate the number of leaves (pieces of data) that lie in the given row or before. For instance, the 8 in the second row shows that there are eight leaves (pieces of data) in the first two rows.

When the row that contains the middle observation(s) is reached, the cumulative frequency is replaced by the number of leaves in that row enclosed by parentheses. Thus, in this case, we see that the middle observations lie in the 210s and that there are 11 data values in the 210s.

The depths following the row that contains the middle observation(s) indicate the number of leaves that lie in the given row or after. In this case, there is only one row following the row in which the middle observations lie, namely, the fourth row. The 1 in the fourth row shows that there is one leaf (piece of data) in the last row. You should compare the stem-and-leaf diagram in Printout 2.6 to Figure 2.7(a) in *IS/6e* and *ES/5e*.

This stem-and-leaf diagram fails to spread the data out very much and hence is only moderately helpful in revealing a pattern. This is because there are too few stems. However, we have the option of changing that with the INCREMENT subcommand. Let us obtain a stem-and-leaf diagram for the cholesterol-level data, using two lines per stem. Using an increment of 5 instead of 10 will force Minitab to use a distance of 5 units between the smallest possible number on one line and the smallest possible number on the next line. Proceed as follows.

Menu commands:

1 Choose Graph ➤ Character Graphs ➤ Stem-and-Leaf... (you may omit the Character Graphs) or Stat ➤ EDA ➤ Stem-and-Leaf...

2 Enter LEVELS in the Variables: text box.

3 Select the Increment: text box and type 5

4 Select [OK]

See Printout 2.7 for the result.

PRINTOUT 2.7
Minitab output for STEM-AND-LEAF using one line per stem

```
Stem-and-leaf of LEVELS  N  = 20
Leaf Unit = 1.0

   1   19 9
   4   20 023
   8   20 7889
 (7)   21 0000234
   5   21 5788
   1   22 1
```

The same result may be obtained from the following session commands.

Session commands:

- Type STEM 'LEVELS'; at the MTB > prompt and press [Enter]
- Type INCREMENT=5. at the SUBC> prompt and press [Enter]

Let us use this printout to discuss depths again. The 8 in the third row of the stem-and-leaf diagram indicates that there are eight leaves (pieces of data) in the first three rows, and the 5 in the fifth row indicates that you will find a total of five leaves (pieces of data) in the last two rows. Compare the stem-and-leaf diagram in Printout 2.7 to the one you find in Example 2.16 of *IS/6e* and *ES/5e*. ■

PROBLEM 2.6 Refer to the age data in Problem 2.3. Apply Minitab to construct a stem-and-leaf diagram,

a) using one line per stem.
b) using two lines per stem. ◇

LESSON 3

Descriptive Measures

GENERAL OBJECTIVE

Besides organizing data into tables and portraying data with graphs, we can also summarize data using descriptive measures. Recall that *descriptive measures* are numbers used to describe data sets. In this lesson we will learn how Minitab can be employed to compute various descriptive measures.

LESSON OUTLINE

3.1 Measures of center

E-3.1:I-3.1

Descriptive measures that indicate where the center or most typical value of a data set lies are called *measures of central tendency* or, more simply, *measures of center.* These are also referred to by many as *averages*. In this section we will see how Minitab can be used to obtain the two most important measures of central tendency, the *mean* and the *median.*

THE MEAN

Recall that the *mean* of a data set is defined to be the sum of the data divided by the number of observations. The Minitab command for computing the mean of a data set is called MEAN.

EXAMPLE 3.1 *Illustrates the MEAN command*

A mathematician spent a summer working for a small mathematical consulting firm. The firm employed a few senior consultants who made between \$800 and \$1050 per week, a few junior consultants who made between \$400 and \$450 per week, and several clerical helpers who made \$300 per week. There was more work in the first half of the summer than in the second half, so there were more employees during the first half. Table 3.1 (WeissStats CD:Tb03-01) and Table 3.2 (WeissStats CD:Tb03-02) give typical lists of weekly earnings for the two halves of the summer measured in dollars. Use Minitab to find the mean of each of the two sets of data.

TABLE 3.1
Data Set I

300	300	300	940	300	300	400
300	400	450	800	450	1050	

TABLE 3.2
Data Set II

300	300	940	450	400
400	300	300	1050	300

SOLUTION For the implementation in this manual, these two sets of data should be located in your worksheet under the names SETI and SETII. Proceed as follows.

Menu commands:

1 Choose Calc ➢ Column Statistics...
2 Select the Mean option button from the Statistic options
3 Select the Input variable: text box and enter SETI
4 Select OK

Session commands:

- Type the command MEAN 'SETI'

The result is displayed in Printout 3.1 on the following page.

PRINTOUT 3.1
Minitab output for MEAN

```
Mean of SETI   =        483.85
```

From the printout you see that the mean for this data set is \$483.85.

The mean of data set `SETII` is determined in an entirely similar manner. The result is shown in Printout 3.2.

PRINTOUT 3.2
Minitab output for MEAN

```
Mean of SETII =        474.00
```

Thus the mean salary of the employees in this data set is \$474.00. ■

PROBLEM 3.1 A college chemistry teacher is concerned about the possible detrimental effects of a poor mathematics background on his students. He takes a sample of 15 students and divides them according to their background in mathematics. Their semester averages turn out to be the following:

Fewer than two years of high school algebra	Two or more years of high school algebra
58	84
81	67
74	65
61	75
64	74
43	92
	83
	52
	81

Use Minitab to determine the mean of each of the two data sets given in the table. Use `LT2` as the name for the data in the first column and `GE2` as the name for the other column of data. ◇

THE MEDIAN

Recall that the **`median`** of a data set is defined to be (1) the data value exactly in the middle of its ordered list, if the number of pieces of data is *odd*, or (2) the average of the two middle data values in its ordered list, if the number of pieces of data is *even*. As you might suspect, the appropriate Minitab command is called MEDIAN.

EXAMPLE 3.2 *Illustrates the MEDIAN command*

Consider again the two sets of salary data in Tables 3.1 and 3.2 on page 44. Employ Minitab to find the median of each of the two data sets.

SOLUTION You have already stored the two data sets in the computer under the names `SETI`, and `SETII`. The execution by menu commands is the same as for the mean, except you select Median from the Statistic field. Similarly, in the session command, type `MEDIAN` instead of `MEAN`.

The results are displayed in Printout 3.3.

PRINTOUT 3.3
Minitab output for MEDIAN

```
Median of SETI  =       400.00
Median of SETII =       350.00
```

Consequently, we see that the median salary of the employees in the first data set is \$400 and that of the employees in the second data set is \$350. ■

PROBLEM 3.2 Refer to Problem 3.1. Use Minitab to find the median of each of the two data sets presented there. ◇

THE MODE

Minitab does not have a MODE command. Nonetheless, we can still use Minitab to obtain the mode of a data set since we can construct a grouped-data table based on single-value grouping.

EXAMPLE 3.3 Illustrates finding the Mode

Consider again the two sets of salary data given in Tables 3.1 and 3.2, entered in the worksheet as `SETI`, and `SETII`. First, we construct a frequency table of just the counts as follows.

Menu commands:

1 Choose Stat ➤ Tables ➤ Tally...
2 Enter both `SETI` and `SETII` in the Variables: text box
3 Select the Counts check box.
4 Select [OK]

The results may be seen in Printout 3.4.

PRINTOUT 3.4
Using TALLY to find the mode

SETI	Count	SETII	Count
300	6	300	5
400	2	400	2
450	2	450	1
800	1	940	1
940	1	1050	1
1050	1	N=	10
N=	13		

The same printout results from the following session commands.

Session commands:

- Type TALLY 'SETI' 'SETII' and press Enter

Observe that you may tally several columns of data at once. Also, observe that you need no subcommands if all you wish to find are the tallies. By inspection of the Count columns, verify that the mode for Data Set I is 300. Also, Data Set II has the same unique mode. ■

PROBLEM 3.3 The U.S. Department of Housing and Urban Development (HUD) compiles data on the number of rooms in occupied housing units. Suppose that a sample of 25 housing units yields the following data on number of rooms:

4	4	6	7	7
8	4	4	8	5
2	9	3	6	7
5	3	5	8	5
6	4	6	5	5

Use Minitab to find the mode of the number of rooms, using HUD as a name for the data. ◇

3.2 The sample mean

E-3.2:I-3.2

The sample mean is just the mean of a set of sample data, which is the sum of the observations divided by n, the sample size. Both of these numbers can be computed directly using Minitab calculator functions.

SUMMATION

Frequently we need to find the sum of a data set. By employing Minitab, we can easily obtain the sum of any data set. The appropriate command is called SUM.

EXAMPLE 3.4 *Illustrates the SUM command*

A student takes four exams in a biology class. His grades are 88, 75, 95, and 100. Find the sum of this grade data with the aid of Minitab.

SOLUTION First, enter the data into a column named EXAMS. Then proceed as follows.

Menu commands:

1 Choose Calc ➤ Column Statistics...

2 Click the Sum option button from the Statistic field

3 Select the Input variable: text box and enter **EXAMS**

4 Select [OK]

Session commands:

- Type the command SUM 'EXAMS' at the MTB > prompt and press [Enter]

The result can be seen in Printout 3.5.

PRINTOUT 3.5
Minitab output for SUM

```
Sum of EXAMS    =      358.00
```

As Printout 3.5 shows, the sum of the exam scores is 358. ■

PROBLEM 3.4 The following table gives the numbers of patients admitted to U.S. hospitals between 1977 and 1982. [Source: American Hospital Association.]

Year	Patients admitted (millions)
1977	37.1
1978	37.2
1979	37.8
1980	38.9
1981	39.2
1982	39.1

Use Minitab to determine the total number of patients admitted to U.S. hospitals between 1977 and 1982, inclusive, using **PATIENTS** for a column name. ◇

THE SAMPLE MEAN

The sample mean is just the mean of a set of sample data, which is the sum divided by n. The value of n may be found in Minitab using the N command. Thus, once data are stored in a column, the value of $\overline{x}$ may be computed using Minitab commands.

EXAMPLE 3.5 *Illustrates the computation of the sample mean*

Each year manufacturers perform mileage tests on new car models and submit the results to the Environmental Protection Agency (EPA). The EPA then tests the vehicles to determine whether the manufacturers are correct.

In 1998, one company reported that a particular model, equipped with a four-speed manual transmission, averaged 29 miles per gallon (mpg) on the highway. Let us suppose that the EPA tested 15 of the cars and obtained the gas mileages given in Table 3.3 on the following page.

TABLE 3.3
Gas mileages

27.3	31.2	29.4	31.6	28.6
30.9	29.7	28.5	27.8	27.3
25.9	28.8	28.9	27.8	27.6

Determine the sample mean of these gas mileages using Minitab commands.

SOLUTION Begin by entering the data in a column named EPA. We need to find the sum of the data their number. Finally, we need to divide these two quantities to compute the sample mean. Proceed as follows.

Menu commands:

1 Choose Calc ➤ Calculator...

2 Type XBAR in the Store results in variable: text box.

3 In the Functions: list box, scroll to the function named Sum and click the mouse to highlight it (Notice that the Select button is activated and the description SUM(number) appears under the button.)

4 Click the Select button to insert SUM(number) in the Expression: text box and observe that number is highlighted

5 Select EPA in the data list box to replace number

6 Click past) in the Expression: text box and click the divide button (the one labeled /) on the displayed calculator

7 Scroll to the function N total in the Functions: menu and select it (Observe that this function is entered as Count (number) in the Expression: text box.)

8 Select EPA once more to replace number

9 Select [OK]

You will discover that the expression Let 'XBAR'=Sum('EPA')/COUNT('EPA') is echoed on the command line in the Session window. This is exactly what the much simpler session command for this execution is.

Session commands:

- Name an empty column XBAR
- Type LET 'XBAR'=SUM('EPA')/COUNT('EPA') at the MTB > prompt and then press [Enter↵]

See Printout 3.6 for the result using the PRINT command.

PRINTOUT 3.6
Using Minitab to find the sample mean

```
MTB > PRINT 'XBAR'
XBAR
  28.7533

MTB > MEAN 'EPA'
  Mean of EPA     =       28.753
```

Of course, to find the sample mean using Minitab, we need only apply the MEAN command and that is a much simpler procedure and far less prone to entry errors. As you see in Printout 3.6, the two results agree except for rounding.† ■

PROBLEM 3.5 The U.S. Federal Bureau of Investigation (FBI) conducts annual surveys to estimate the average value lost by victims of selected crimes. Suppose that a sample of shoplifting offenses yields the following data on value lost

$88	70	121	159
41	39	82	107
69	53	18	95

Enter the data in a column named FBI and determine the sample mean of these losses with the aid of Minitab using the commands SUM and N. Continue to use the column named XBAR. Compare your answer with the result of applying the MEAN command. ◇

3.3 Measures of variation; the sample standard deviation

E-3.3:I-3.3

Descriptive measures that indicate the amount of variation in a data set are called *measures of variation* or *measures of spread.* We will learn in this section how Minitab can be applied to obtain two of the most frequently used measures of spread, the *range* and the *standard deviation.*

THE RANGE

The *range* of a data set is the difference between the largest and smallest data values in the data set:

$$\text{Range} = \text{Largest value} - \text{Smallest value}.$$

Minitab has a RANGE command that will compute the value of the range of a set of data directly.

EXAMPLE 3.6 *Illustrates how to find the range using Minitab*

The heights, in inches, of the five starting players on a basketball team are 67, 72, 76, 76, and 84. Use Minitab to obtain the range of these height data.

† Incidentally, the argument number in the menu commands is a misnomer in a way; it would be better if it read column.

SOLUTION First, store the height data in a column named `HEIGHT`. Then proceed as follows.

Menu commands:

1 Choose Calc > Column Statistics...
2 Select Range from among the Statistic option buttons
3 Select the Input variable: text box and enter `HEIGHT`
4 Select [OK]

Session commands:

- Type `Range 'HEIGHT'` at the `MTB >` prompt and press [Enter]

The result is shown in Printout 3.7.

PRINTOUT 3.7 Minitab output for the range

```
Range of HEIGHT   =        17.000
```

From the printout, you see the range of heights is 17 inches. ■

PROBLEM 3.6 Repeat Example 3.6 for the data of Problem 3.5. Apply the appropriate Minitab commands to find the range of the losses. ◇

THE SAMPLE STANDARD DEVIATION

Recall that the *sample standard deviation*, s, of n pieces of sample data is defined by

$$s = \sqrt{\frac{\Sigma(x - \overline{x})^2}{n - 1}}$$

and can also be computed by using the shortcut formula

$$s = \sqrt{\frac{\Sigma x^2 - (\Sigma x)^2/n}{n - 1}}.$$

This last formula is quite suitable for use by calculating devices that are capable of storing three quantities from the data. These three quantities are the sample size, n; the sum, Σx; and the sum of squares, Σx^2. Most of the hand-held calculators on the market today have this capability.

Regardless of which formula is used to compute s, the calculations are generally tedious and time-consuming. We can avoid doing the calculations by employing Minitab's STDEV command.

EXAMPLE 3.7 *Illustrates the STDEV command*

Consider again the heights, in inches, of the five starting players on a team as discussed in Example 3.7. The heights are 67, 72, 76, 76, and 84. Use Minitab to determine the sample standard deviation of these heights.

SOLUTION In Example 3.6 we stored the heights in a column named HEIGHT. To obtain the sample standard deviation of those heights, we proceed as follows.

Menu commands:

1 Choose Calc ➤ Column Statistics... and select Standard deviation from among the Statistic option buttons

2 Select the Input variable: text box and enter HEIGHT

3 Select [OK]

Session commands:

- Type STDEV 'HEIGHT' at the MTB > command and press [Enter↵]

The result is shown in Printout 3.8.

PRINTOUT 3.8
Minitab output for STDEV command

```
Standard deviation of HEIGHT   =       6.2450
```

Thus the sample standard deviation of the heights of the five starting players is $s = 6.2450$ inches. ■

PROBLEM 3.7 Determine the sample standard deviation of the loss data given in Problem 3.5 with the aid of Minitab. ◇

3.4 The five-number summary; boxplots

E-3.4:I-3.4

Minitab has a command called DESCRIBE that will display several descriptive measures of a data set simultaneously. Among these computed and displayed by this command are the mean and the median of the data set, as well as the first and third quartiles.

QUARTILES; INTERQUARTILE RANGE

The DESCRIBE command will calculate and print quartiles for any data set. It is pointed out just before Example 3.15 in *IS/6e* and *ES/5e*, that not all statisticians define quartiles in exactly the same way. While these other definitions may lead to slightly different values, the differences tend to be unimportant in practice.

EXAMPLE 3.8 *Illustrates the use of the DESCRIBE command*

The A. C. Nielsen Company publishes data on the TV viewing habits of Americans by various characteristics in *Nielsen Report on Television.* A sample of 20 people yields the weekly viewing times, in hours, is considered in Example 3.15 of *IS/6e* and *ES/5e*. It displayed here as Table 3.4 (WeissStats CD:Tb03-13).

TABLE 3.4 Weekly TV viewing times

25	41	27	32	43
66	35	31	15	5
34	26	32	38	16
30	38	30	20	21

Use Minitab to locate the first and third quartiles for this data set.

SOLUTION Be sure the data are stored in your worksheet under the name TIMES. Proceed as follows.

Menu commands:

1 Choose Stat ➤ Basic Statistics ➤ Display Descriptive Statistics...

2 Enter TIMES in the Variables: text box.

3 Select [OK]

Session commands:

- Type DESCRIBE 'TIMES' at the MTB > prompt and press [Enter↵]

See Printout 3.9 for a display of the output.

PRINTOUT 3.9 Minitab output for DESCRIBE

```
Variable          N      Mean    Median    TrMean     StDev    SEMean
TIMES            20     30.25     30.50     29.67     12.65      2.83

Variable        Min       Max        Q1        Q3
TIMES          5.00     66.00     22.00     37.25
```

The second row of entries of the output display the smallest data value under Min and the largest data value under Max. You may verify that the difference between these values produces the range as discussed on page 50 of this manual. The final two entries give the first quartile under Q1 and the third quartile under Q3. Note that the values given are different than the ones published in Example 3.17 of *IS/6e* and *ES/5e*. Here, IQR has the value 15.25. ■

PROBLEM 3.8 The U.S. Energy Information Administration publishes figures on residential energy consumption and expenditures. Suppose that a sample of 36 households using electricity as their primary energy source yields the following data on last year's energy expenditures measured in dollars.

1376	1452	1235	1480	1185	1327
1059	1400	1227	1102	1168	1070
1180	1221	1351	1014	1461	1102
976	1394	1379	987	1002	1532
1450	1177	1150	1352	1266	1109
949	1351	1259	1179	1393	1456

Enter the data in a column named ENERGY and use the DESCRIBE to find the first and third quartiles. Compute the value of the interquartile range. ◇

BOXPLOTS

We can use Minitab to obtain a boxplot, or box-and-whisker diagram, using the BOXPLOT command. We will illustrate the use of that command and the meaning of the output by applying it to the TV viewing-time data previously considered.

EXAMPLE 3.9 Illustrates the BOXPLOT command

Use Minitab to obtain a boxplot for the TV viewing times recorded in Table 3.4 of Example 3.8 just discussed.

SOLUTION Assume the data are still stored the data under the name TIMES. Proceed as follows.

Menu commands:

1 Choose Graph ➤ Boxplot... and enter TIMES in the Graph Variables: text box under Y for Graph 1

2 Select OK twice

Session commands:

- Type BOXPLOT 'TIMES'; at the MTB > prompt and press Enter

FIGURE 3.1 Minitab boxplot of TIMES data

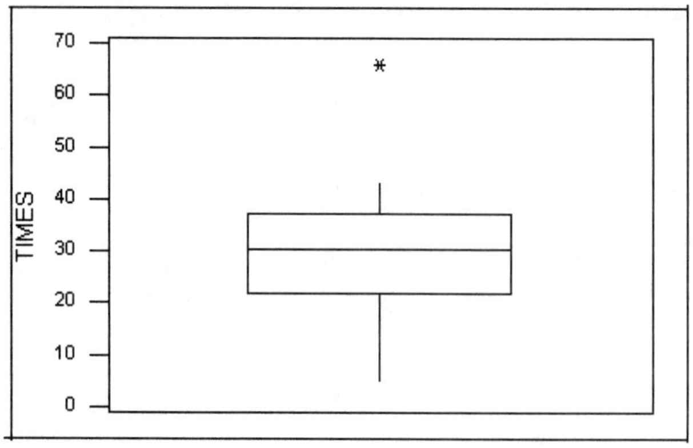

The output is displayed in Figure 3.1.

The graph is plotted vertically instead of horizontally as in your textbook, but it conveys exactly the same information.†

The boxplot consists of, first, a box inserted between Q_1 and Q_3 on a vertical scale of measurement for the `TIMES` data, between 22.0 and 37.25 in this example. A horizontal line is then drawn in the box at the median, 30.5 in this case, and symmetry or lack of it is indicated by how close to the middle of the box this median line is located.

Next, some reasonable limits within which you would expect to find data of this type is determined by the interquartile range, 15.25 for our example. A lower limit is determined by $Q_1 - 1.5 * IQR = -0.875$ and $Q_3 + 1.5 * IQR = 60.125$ is the upper limit. Next, *adjacent values* are defined as the lowest and highest observations within these limits; in our example, these numbers are the minimum value of 5 and the second to the largest value 43. Two lines, called *whiskers*, are drawn from these values to the box—hence the name box-and-whisker diagram.

Had the maximum value of 66 been below the upper limit, all of the data would have been included between the limits. As it turns out, 66 is greater than the upper limit and so it is classified by Minitab as an *outlier* and marked with an asterisk (*). As discussed in *IS/6e* , this does not mean that 66 could not have occurred with this data set, nor should it be discarded out of hand. A better term for this extreme value would be *potential outlier*. As a diagnostic tool, the boxplot helps the experimenter obtain a visual impression of the data and other considerations should be brought to bear on handling potential outliers. ■

PROBLEM 3.9 Refer to Problem 3.8. Repeat the details of Example 3.9 to analyze the data using the DESCRIBE and BOXPLOT commands. ◇

3.5 Descriptive measures for populations **E-3.5:I-3.5**

We will now see how Minitab can be used to determine the mean and standard deviation of a finite population. Those two parameters are the most important descriptive measures of a population.

THE POPULATION MEAN

The *population mean*, μ, of a finite population of size N is defined by

$$\mu = \frac{\Sigma x}{N}.$$

Thus we see that the population mean is just the mean of a set of population data. Consequently, to determine a population mean using Minitab, we simply employ the MEAN command, as indicated in the following example.

† You may effect a horizontal plot by adding the subcommand `TRANSPOSE` to the session command or electing it as an option in the menu command.

EXAMPLE 3.10 Illustrates finding the population mean

Table 3.15 in *IS/6e* and *ES/5e* presents some weights for the 20 active players on a soccer team. The weights of these players are summarized here in Table 3.5. Determine the population mean weight of these 20 players with the aid of Minitab.

TABLE 3.5
U.S. Women's World /Cup soccer team, 1999

145	125	140	125	120
130	130	130	130	150
130	145	130	130	125
125	165	135	145	140

SOLUTION The weight data (WeissStats CD:Tb03-15) should be stored in a column named `WEIGHT` in your worksheet. Proceed as follows.

Menu commands:

1 Choose Calc ➤ Column Statistics...
2 Select the Mean radio button from among the Statistic options
3 Select the Input variable: text box and enter `WEIGHT`
4 Select [OK]

Session commands:

■ Type `MEAN 'WEIGHT'` at the `MTB >` prompt and press [Enter↵]

Printout 3.10 shows the result.

PRINTOUT 3.10
Population mean using Minitab

```
Mean of WEIGHT = 134.75
```

Hence the population mean weight of the players is $\mu = 134.75$ lb. Nothing new here and only the notation μ tells us that the computation represents that of a population mean and not a sample mean. ■

PROBLEM 3.10

The corresponding heights, in inches, for the players in Example 3.10 are given in the following table.

68	63	66	67	64
67	65	65	65	70
66	71	64	65	66
62	71	69	67	69

Apply Minitab to obtain the population mean height of these 20 players. The data stored in a column named `HTS` ◇

THE POPULATION STANDARD DEVIATION

Recall that the *population standard deviation*, σ, of a finite population of size N is defined by

$$\sigma = \sqrt{\frac{\Sigma(x-\mu)^2}{N}}.$$

The essential difference in *computation* between the population standard deviation and the sample standard deviation is we divide by *one less* than the number of pieces of data in the formula for the sample standard deviation.

Minitab does not have a command to obtain σ, the population standard deviation. However, we can adjust for the fact that STDEV divides by one less than the number of pieces of data instead of dividing by the number of pieces of data. Thus the standard deviation, σ, of a population can be obtained from the formula

$$\sigma = \sqrt{\frac{N-1}{N}} * \text{STDEV}.$$

EXAMPLE 3.11 *Illustrates modifying the STDEV command*

Consider again the weights of members of the U.S. Women's soccer team, given in Table 3.5. Apply Minitab to determine the population standard deviation of the weights.

SOLUTION We have already stored the weight data in `WEIGHT`. Using the formula discussed directly before this illustration, we can determine σ as follows in the session window.

Session commands:

- Type `LET K7=SQRT((N('WEIGHT')-1)/N('WEIGHT'))*STDEV('WEIGHT')` at the `MTB >` prompt and press Enter

See Printout 3.11 for a display of the result using the PRINT command.

PRINTOUT 3.11
Modifying STDEV

```
MTB > PRINT K7
K7     10.6624
```

Consequently, we see that the population standard deviation of the weights is $\sigma = 10.6624$ lb. ■

It is easy to create a local macro to automate this adjustment so that σ can be computed for any named column in the workspace. Use a text editor to save the lines in Printout 3.12 on the following page in a file named SIG.MAC.

PRINTOUT 3.12
Local Macro for σ

```
MACRO
SIG X SIGMA
MCONSTANT N SIGMA
MCOLUMN X
LET N=COUNT(X)
LET SIGMA= SQRT((N-1)/N)* STDEV(X)
PRINT SIGMA
ENDMACRO
```

Save this SIG.MAC file in the folder where the rest of your macros are located (typically in MACROS).

EXAMPLE 3.12 Illustrates the use of the % macro SIG

Consider again the weights for the 20 active players on a soccer team as in the last example. Use the %macro SIG to determine the population standard deviation of the weights.

SOLUTION We have already stored the weight data in WEIGHT. To find σ for the particular column of data WEIGHT, you have merely to type the following session command.

Session commands:

- Type %SIG 'WEIGHT' 'SIGMA' at the MTB > prompt and press Enter

The value of the population standard deviation will appear in the session window as exemplified in Printout 3.13.

PRINTOUT 3.13
Applying the SIG macro to 'WEIGHT'

```
Executing from file: SIG.MAC

Data Display

SIGMA    10.6624
```

Observe that the result is exactly the same as that computed by hand. This is not surprising since we placed those very steps in the macro. ■

PROBLEM 3.11

Refer to Problem 3.11. Use the % macro SIG to determine the population standard deviation of the heights of the 20 players. ◇

LESSON 4

Probability Concepts

GENERAL OBJECTIVE

In Lessons 2 and 3, we learned how Minitab can be used to accomplish the tasks of descriptive statistics. Now we will see how Minitab can be applied to perform some of the probability computations discussed in Chapter 4 of *IS/6e* and Chapter 5 of *ES/5e* .

LESSON OUTLINE

4.1 Equally likely outcomes

E-5.1:I-4.1

When the possible outcomes of an experiment are *equally likely* to occur, probabilities are relative frequencies or percentages. Specifically, suppose there are N equally likely outcomes for an experiment. Then the probability an event occurs equals the number of ways, f, that the event can occur, divided by the total number, N, of possible outcomes. That is, the probability equals

Number of ways event can occur

$$\frac{f}{N}.$$

Total number of possible outcomes

We can use Minitab to perform the calculations required in applying the f/N rule, as the following example shows.

EXAMPLE 4.1 *Illustrates how Minitab can be used to apply the f/N rule*

The U.S. Bureau of the Census compiles data on family income and publishes its findings in *Current Population Reports*. Table 4.1 gives a frequency distribution of annual income for U.S. families in 1998.

TABLE 4.1 Frequency distribution of annual incomes for U.S. families

Income	Frequency
Under $10,000	4.187
$10,000–$14,999	3,653
$15,000–$24,999	8,639
$25,000–$34,999	8.991
$35,000–$49,999	12,192
$50,000–$74,999	15,676
$75,000 & over	18,192
	71,535

Use Minitab to find the probability that a randomly selected family makes
a) between $50,000 and $74,999 inclusive.
b) between $25,000 and $74,999 inclusive.

SOLUTION First, store the frequencies given in the second column of Table 4.1 into a column named `f`. The total number of families in the city, N, equals the sum of the frequencies in the second column of Table 4.1. Since the frequencies are stored in `f`, we can find N with Minitab by using the SUM command.

a) To determine the probability that a randomly selected family makes between $50,000 and $74,999, we need to divide the number of such families by the total number of families in the city. Now, the number of families, f, with incomes between $10,000 and $74,999 is stored in the sixth row of the Frequency column (see Table 4.1). Proceed as follows.

Session commands:

- Type LET K1='f'(6)/SUM('f')

This much stores the desired probability in K1. Typing PRINT K1 instructs Minitab to have the computer display that probability. See Printout 4.1.

PRINTOUT 4.1 Computing probabilities in Minitab

```
MTB > PRINT K1
K1    0.219137
```

Thus the probability that a randomly selected family makes between $50,000 and $74,999 is 0.219137, or about 22%.

b) For this part, we want to find the probability that a randomly selected family makes between $25,000 and $74,999. The number of such families is the sum of the fourth, fifth, and sixth rows of f (see Table 4.1). That sum can be obtained in several ways. Here is one.

Menu commands:

1 Choose Manip ➤ Copy Columns...
2 Enter f in the Copy from columns: text box
3 Select the To columns: text box and type RESULT
4 Select the Use Rows... button to open the Copy Columns-Use Rows dialog box
5 Select the Use rows (e.g., 1:4 12): option button
6 Press Tab and type 4:6 in the text box
7 Select OK twice

Session commands:

- Name an empty column RESULT
- Type Copy 'f' 'RESULT'; at the MTB > prompt and press Enter
- Type Use 4:6. at the SUBC> prompt and press Enter

This copies rows 4 through 6 (the colon stands for "through") of f into RESULT.

Now the sum of the fourth, fifth, and sixth rows of f can be obtained by applying the SUM command to RESULT. Type LET K2=SUM('RESULT')/SUM('f') and the desired probability will be stored in K2. See Printout 4.2 for output.

PRINTOUT 4.2 Adding probabilities in Minitab

```
MTB > PRINT K2
K2    0.515328
```

As the printout reveals, about 52% of U.S. families made between \$25,000 and \$74,999 in 1995. ■

PROBLEM 4.1 Apply Minitab to the data in Table 4.1 to find the probability that a randomly selected family makes
a) less than \$15,000.
b) between \$10,000 and \$34,999.
c) either less than \$10,000 or at least \$75,000. ◇

4.2 Contingency tables; joint distributions

I-4.4

Contingency (or two-way) tables, are used to display frequency distributions for bivariate data for a population. From such a table we can construct various probability distributions and Minitab can assist us in that matter. The data displayed in Table 4.2 are adapted from the *Arizona State University Statistical Summary* (see Table 4.6 in Example 4.14 of *IS/6e*).

The table provides us with a frequency distribution obtained by cross classifying the faculty according to the two characteristics age and rank. In this section we will learn how Minitab can be applied to store contingency tables in the computer and to manipulate such tables to obtain joint probability distributions.

TABLE 4.2 Contingency table for age vs. rank of faculty members

AGE \ RANK	Full Professor R_1	Associate Professor R_2	Assistant Professor R_3	Instructor R_4	Total
Under 30 A_1	2	3	57	6	68
30–39 A_2	52	170	163	17	402
40–49 A_3	156	125	61	6	348
50–59 A_4	145	68	36	4	253
60 & over A_5	75	15	3	0	93
Total	430	381	320	33	1164

We will use Table 4.2 as a reference for several examples in this chapter.

CONTINGENCY TABLES IN MINITAB

First, let us examine Minitab's version of a contingency table.

EXAMPLE 4.2 *Minitab contingency table*

Use Minitab to store the contingency table, Table 4.2, into the computer. After the table is stored, have Minitab display the contingency table.

SOLUTION First, name three empty, neighboring columns AGEC,RANK,FREQ in a Minitab worksheet. There are 20 cells (ignoring the Total row and column) containing frequencies for this table. Enter the data in triplets under the appropriate columns. Thus, for example, for the upper left hand corner of Table 4.2, enter (Under 30, Full, 3), corresponding to the number of Full Professors who are under 30 years of age; follow this with the triplet (Under 30, Assoc, 3), then (Under 30, Asst, 57), followed by (Under 30, Instr, 6) to complete the entry of the first row in the table. The order of entry of subsequent triplets is immaterial as long as each unique triplet is entered. However, you may find it easier if you proceed row by row or column by column. When you complete the entry of 20 cells, proceed as follows to display the contingency table.

Menu commands:

1. Choose Editor ➤ Column ➤ Value order... and click the radio button next to Order of occurrence in worksheet
2. Select Stat ➤ Tables ➤ Cross Tabulation...
3. Enter AGEC and RANK in the Classification variables: text box
4. Select the Counts check box under Display and deselect (clear) all other check boxes
5. Select the Frequencies are in: text box
6. Press [Tab] and enter FREQ
7. Select [OK]

Session commands:

- Type Table 'AGEC''RANK'; at the MTB > prompt and press [Enter]
- Type Frequencies 'FREQ'; at the SUBC> prompt and press [Enter]
- Type Counts; at the SUBC> prompt and press [Enter]
- Type Layout 1 1.

Minitab will display the contingency table that appears in Printout 4.3 on the following page. Compare this with Table 4.2 and you will see that the entries are identical.† You may also wish to compare this with Table 4.6 in *IS/6e* to see that the entries agree.

† The only reason for the Layout option was to be sure that the tables matched; without this, Minitab simply orders the two classifications alphabetically.

PRINTOUT 4.3
Displaying a Minitab contingency table

```
Rows: AGEC      Columns: RANK

            Full   Assoc    Asst   Instr     All

Under 30       2       3      57       6      68
30-39         52     170     163      17     402
40-49        156     125      61       6     348
50-59        145      68      36       4     253
60&over       75      15       3       0      93
All          430     381     320      33    1164

 Cell Contents --
                Count
```

Totals are identified by the name `All` in the Minitab printout. Thus, there are a total of 402 faculty members that are age 30–39. Of these, 170 have the rank of Associate Professor. Of the 320 Assistant Professors, most of them, 163, are between the ages of 30–39. No instructor at this institution is 60 years of age or over. ■

PROBLEM 4.2 The following contingency table was obtained by cross-classifying the members of the U.S. House of Representatives (98th Congress) according to years of seniority and political party. [Source: U.S. Congress, Joint Committee on Printing.]

YEARS	PARTY: Democrat P_1	PARTY: Republican P_2	Total
Under 2 S_1	59	25	84
2–9 S_2	134	93	227
10–19 S_3	44	41	85
20–29 S_4	20	8	28
30 & over S_5	10	0	10
Total	267	167	434

Employ Minitab to store the display the contingency table. Use the name `NUMBER` to store the frequencies in each cell. ◇

JOINT PROBABILITY DISTRIBUTIONS; MARGINAL DISTRIBUTIONS

Once we have a contingency table stored, it is easy to obtain the joint probability distribution corresponding to the contingency table. Let us return to the situation of Example 4.2.

EXAMPLE 4.3 *Illustrates how to obtain joint probability distributions*

Recall that Table 4.2 provides a contingency table that cross-classifies the faculty at Arizona State University by age and rank. Suppose a faculty member is selected at random. Apply Minitab to obtain the joint probability distribution for the age versus rank data.

SOLUTION As we see from Table 4.2, there is a total of $N = 1164$ faculty members. Thus, the joint probability distribution for the age versus rank data is obtained by dividing each frequency in Table 4.2 by 1164 according to the f/N rule. Minitab will do this automatically now that we have the table stored. Proceed as follows.

Menu commands:

1 Select Stat ➤ Tables ➤ Cross Tabulation...
2 Enter **AGEC** and **RANK** in the Classification variables: text box
3 Select the Total percents check box under Display and deselect (clear) all other check boxes
4 Select the Frequencies are in: check box
5 Press Tab and enter FREQ
6 Select OK

The result may be seen in Printout 4.4.

PRINTOUT 4.4 Displaying a joint probability distribution in Minitab

```
Rows: AGEC      Columns: RANK

              Full     Assoc      Asst     Instr       All

Under 30      0.17      0.26      4.90      0.52      5.84
30-39         4.47     14.60     14.00      1.46     34.54
40-49        13.40     10.74      5.24      0.52     29.90
50-59        12.46      5.84      3.09      0.34     21.74
60&over       6.44      1.29      0.26        --      7.99
All          36.94     32.73     27.49      2.84    100.00

 Cell Contents --
                    % of Tbl
```

The same printout results from the following session commands.

Session commands:

- Type `Table 'AGEC''RANK';` at the `MTB >` prompt and press Enter
- Type `Frequencies 'FREQ';` at the `SUBC>` prompt and press Enter
- Type `Totpercents;` at the `SUBC>` prompt and press Enter
- Type `Layout 1 1.` at the `SUBC>` prompt and press Enter

The entries in this table are given as percents, consistent with our interpretation of probabilities found by the f/N rule. Thus, 5.84% of the faculty Associate Professors AND are age 50–59. The probability of drawing a Full Professor at random from this population is 0.3694 or 36.94% as seen in the first column total in the ALL row. As noted in *IS/6e* , this probability is one item in the marginal distribution corresponding to this row. Similarly, the probability of drawing a professor at random who is age 50–59 from this population is 0.2174 or 21.74% as seen in the fourth row total in the ALL column. You should compare the joint probability distribution in Printout 4.4 to the one given in Table 4.7 of *IS/6e* . Except for rounding, they are the same. ■

PROBLEM 4.3 Refer to the contingency table in Problem 4.2. Suppose a representative is selected at random. Use Minitab to find the joint probability distribution for the years of seniority versus political party data. What is the marginal probability of obtaining a congressman age 20–29? ◇

4.3 Conditional probability

I-4.5

In Example 4.17 of *IS/6e* , you learned how to find conditional probabilities from a contingency table. Minitab's TABLE command allows you to store all such conditional probabilities simultaneously once you have the contingency table stored. Let us illustrate with the next example.

EXAMPLE 4.4 ***Illustrates how to obtain conditional probability distributions using Minitab***

Determine the conditional probabilities corresponding to each Rank classification with RANK the conditioning variable using Minitab.

SOLUTION With the table still in the worksheet (re-enter it if necessary) proceed as follows.

Menu commands:

1. Select Stat ➤ Tables ➤ Cross Tabulation...
2. Enter AGEC and RANK in the Classification variables: text box
3. Select the Column percents check box under Display and deselect (clear) all other check boxes

4 Select the Frequencies are in: check box

5 Press Tab and enter FREQ

6 Select OK

Session commands:

- Type Table 'AGEC''RANK'; at the MTB > prompt and press Enter
- Type Frequencies 'FREQ'; at the SUBC> prompt and press Enter
- Type Colpercents; at the SUBC> prompt and press Enter
- Type Layout 1 1. at the SUBC> prompt and press Enter

The result may be seen in Printout 4.5

PRINTOUT 4.5 Conditional probabilities conditioned on RANK in Minitab

```
Rows: AGEC     Columns: RANK

              Full    Assoc     Asst    Instr      All

Under 30      0.47     0.79    17.81    18.18     5.84
30-39        12.09    44.62    50.94    51.52    34.54
40-49        36.28    32.81    19.06    18.18    29.90
50-59        33.72    17.85    11.25    12.12    21.74
60&over      17.44     3.94     0.94       --     7.99
All         100.00   100.00   100.00   100.00   100.00

 Cell Contents --
                % of Col
```

The entries in this table are given in percents as before. Thus, we can read the conditional probability that a faculty member is in the age 50–59 bracket given that he or she is an Assistant professor directly from the table as 0.1125. This is consistent with the computations given in *IS/6e* . Here all such conditional probabilities may be read from the printout directly. The marginal probabilities are maintained in the ALL column. Thus, the probability of a faculty member age 50–59 regardless of rank is 0.2174 ■

Similarly of course all conditional probabilities conditioned on the AGE variable may be obtained as illustrated in the next example.

EXAMPLE 4.5 *Illustrates how to obtain conditional probability distributions using Minitab*

Determine the conditional probabilities corresponding to each Rank classification with AGE the conditioning variable using Minitab.

SOLUTION With the table still in the worksheet (re-enter it if necessary) proceed as follows.

Menu commands:

1 Select Stat ➤ Tables ➤ Cross Tabulation...
2 Enter AGEC and RANK in the Classification variables: text box
3 Select the Row percents check box under Display and deselect (clear) all other check boxes
4 Select the Frequencies are in: text box
5 Press Tab and enter FREQ
6 Select OK

Session commands:

- Type Table 'AGEC''RANK'; at the MTB > prompt and press Enter
- Type Frequencies 'FREQ'; at the SUBC> prompt and press Enter
- Type Rowpercents; at the SUBC> prompt and press Enter
- Type Layout 1 1. at the SUBC> prompt and press Enter

The result may be seen in Printout 4.6

PRINTOUT 4.6 Conditional probabilities conditioned on AGE in Minitab

```
 Rows: AGEC     Columns: RANK

             Full     Assoc      Asst     Instr       All

Under 30     2.94      4.41     83.82      8.82    100.00
30-39       12.94     42.29     40.55      4.23    100.00
40-49       44.83     35.92     17.53      1.72    100.00
50-59       57.31     26.88     14.23      1.58    100.00
60&over     80.65     16.13      3.23        --    100.00
All         36.94     32.73     27.49      2.84    100.00

 Cell Contents --
                 % of Row
```

The entries in this table are given as percents as before. Thus, we can read the conditional probability that a faculty member is an Instructor given that he or she is in the age 30–39 age bracket given that he or she is an Assistant professor directly from the table as 0.0423. All such conditional probabilities may be read from the printout directly. ■

PROBLEM 4.4 Refer to the contingency table in Problem 4.2. Use Minitab to determine the conditional probability of a Republican among congressman who have been in office under 2 years. Find the conditional probability that a congressman has been in office under 2 years given that he or she is a Democrat. ◇

LESSON 5

Discrete Random Variables

GENERAL OBJECTIVE

Minitab is also useful for dealing with random variables and probability distributions. Recall that a *random variable* is a numerical quantity whose value depends on chance. In this lesson we will learn how Minitab can be used to perform many of the computations required for *discrete* random variables.

LESSON OUTLINE

5.1 Discrete random variables and probability distributions

E-5.4:I-5.1

A *discrete random variable* is a random variable whose possible values form a discrete data set. Generally, a discrete random variable involves a *count* of something. Examples of discrete random variables are the number of cars owned by a randomly selected family, the number of people waiting for a haircut in a barber shop, and the number of households in a sample that own a color television set.

The *probability distribution* of a discrete random variable consists of a listing of (or formula for) the probabilities associated with the various values of the random variable. Often, the probability distribution of a discrete random variable is obtained from a frequency distribution. For such cases, Minitab can be of help in performing the necessary computations.

EXAMPLE 5.1 *Illustrates how Minitab can be used to obtain probability distributions*

Table 5.1 gives a frequency distribution for the enrollment by grade level in public elementary schools in the U.S. [Source: U.S. National Center for Education Statistics]. (0 = kindergarten, 1 = first grade, and so on). Frequencies are in thousands of students.

TABLE 5.1 Frequency distribution for enrollment by grade in U.S. public elementary schools

Grade level y	Frequency f
0	4,043
1	3,593
2	3,440
3	3,439
4	3,426
5	3,372
6	3,381
7	3,404
8	3,302

Suppose a student in elementary school is to be selected at random. Let Y denote the grade of the student obtained. Employ Minitab to determine the probability distribution of the random variable Y.

SOLUTION To obtain the probability distribution of Y we need to apply the f/N rule to each of the frequencies in the second column of Table 5.1. First, store the frequency distribution in Table 5.1 into two columns named `y` and `fy` in the usual manner and proceed as follows.

Menu commands:

1 Choose Calc ➤ Calculator...
2 Type P(Y=y) in the Store results in variable: text box
3 Press [Tab] and type fy/SUM(fy) in the Expression: box
4 Select [OK]

Session commands:

- Name an empty column P(Y=y) if necessary
- Type Let P(Y=y)=fy/SUM(fy) at the MTB > prompt and press [Enter]

To display the results proceed as follows.

Menu commands:

1 Choose Manip ➤ Display data...
2 Enter y and P(Y=y) in the Columns, constants, and matrices to display: text box
3 Select [OK]

Session commands:

- Type PRINT 'y' 'P(Y=y)' at the MTB > prompt and press [Enter]

The results should appear as in Printout 5.1.

PRINTOUT 5.1 Probability distribution of Y

Row	y	P(Y=y)
1	0	0.128758
2	1	0.114427
3	2	0.109554
4	3	0.109522
5	4	0.109108
6	5	0.107389
7	6	0.107675
8	7	0.108408
9	8	0.105159

You should compare the probability distribution in Printout 5.1 to the ones you will find in *IS/6e* and *ES/5e*. With such a probability distribution at hand, it is a simple matter to determine any probability involving the random variable Y. Thus the probability that the grade level of a student selected at random from this distribution being either 0 or 1 is the sum 0.129+0.114=0.243. ■

PROBLEM 5.1 According to the U.S. Bureau of the Census, a frequency distribution for the number of persons per household in the U.S. is as shown in the table on the following page. Frequencies are in millions.

Number of persons z	Number of households f
1	19.4
2	26.5
3	14.6
4	12.9
5	6.1
6	2.5
7	1.6

Suppose a U.S. household is to be selected at random. Let Z denote the number of persons in the household chosen. Employ Minitab to obtain the probability distribution of the random variable Z. Use `z`, `fz` and `P(Z=z)` to store the data. ◇

INTERPRETATION OF PROBABILITY DISTRIBUTIONS

The relative frequency interpretation of probability implies that, in a large number of independent repetitions of the experiment, the relative frequency of occurrence of an event will be close to the probability of that event. We can use Minitab to mimic independent repetitions of an experiment involving the observation of a random variable using a special command called RANDOM. This process is referred to as *simulation* and is a very powerful tool using random number generators. In contrast to the SIMPLE command that you studied in Lesson 1, the RANDOM command allows sampling with replacement, making each trial independent of the others.

Among other subcommands for the RANDOM command is one called DISCRETE. This subcommand allows us to simulate observations from any discrete distribution stored in a pair of columns in the Minitab workspace. We will illustrate this with the next example.

EXAMPLE 5.2 Illustrates the RANDOM command and the DISCRETE subcommand

In Table 5.2 that follows, we display the probability distribution for X, the number of heads obtained in three tosses of a balanced dime, duplicating Table 5.6 in *IS/6e* (Table 5.11 in *ES/5e*). Use Minitab to simulate 1000 observations of the number of heads obtained in three tosses of a balanced dime and compare the relative frequencies with the probabilities.

TABLE 5.2 Probability distribution of the random variable X, the number of heads obtained in three tosses of a balanced dime

No. of heads x	Probability $P(X = x)$
0	0.125
1	0.375
2	0.375
3	0.125

SOLUTION First, store the probability distribution in two columns named x and Px in the usual manner. The simulation may be carried out as follows.

Menu commands:

1 Choose Calc ➤ Random Data ➤ Discrete...
2 Type 1000 in the Generate [ ] rows of data text box
3 Select the Store in column(s): text box and type NUMHEADS
4 Select the Values in: text box and enter x
5 Select the Probabilities in: text box and enter Px
6 Select [OK]

Session commands:

- Name any empty column NUMHEADS if necessary
- Type RANDOM 1000 'NUMHEADS'; at the MTB > prompt and press [Enter]
- Type DISCRETE 'x''Px'. at the SUBC> prompt and press [Enter]

This much collects 1000 independent observations of this random variable and stores them in the column NUMHEADS. We need to process these results to discover how many times each observation occurred and the relative frequency of occurrence. We can do this with the TALLY command, expressing relative frequency as a percentage as follows.

Menu commands:

1 Choose Stat ➤ Tables ➤ Tally...
2 Enter NUMHEADS in the Variables: text box
3 Select the Counts and Percents check boxes from Display check boxes
4 Select [OK]

Session commands:

- Type TALLY 'NUMHEADS'; at the MTB > prompt and press [Enter]
- Type COUNTS; at the SUBC> prompt and press [Enter]
- Type PERCENTS. at the SUBC> prompt and press [Enter]

The results may be viewed in Printout 5.2 and will very likely differ from the one you obtain.

PRINTOUT 5.2 Minitab output for the TALLY command

```
NUMHEADS  Count  Percent
       0    121    12.10
       1    397    39.70
       2    344    34.40
       3    138    13.80
      N=   1000
```

Notice the somewhat close agreement between the percentages obtained and the probabilities listed in Table 5.2, expressed as percentages. ■

PROBLEM 5.2

Using the commands illustrated in Example 5.2, generate your own sample of 1000 independent observations of the distribution in Table 5.2. Tally the results and compare the relative frequencies to the probabilities. ◇

5.2 Mean and standard deviation of a discrete random variable

E-5.5:I-5.2

In this section we will learn how Minitab can be used to determine the mean and standard deviation of a discrete random variable. We first consider the mean of a discrete random variable. No menu commands are offered in this section.

MEAN OF A DISCRETE RANDOM VARIABLE

The *mean* μ of a discrete random variable X is defined by

$$\mu = \Sigma x P(X = x)$$

We can use Minitab to compute the mean of a discrete random variable, as illustrated in the next example.

EXAMPLE 5.3 *Illustrates how to use Minitab to compute the mean of a discrete random variable*

Prescott National Bank has six tellers available to serve customers. The number of tellers busy with customers at, say, 1:00 P.M. varies from day to day and depends on chance; so it is random variable, which we will call X. Past records indicate that the probability distribution of X is as shown in the Table 5.3. Employ Minitab to determine the mean, μ, of the random variable X.

TABLE 5.3 Probability distribution of the random variable X, the number of Tellers busy with customers

x	$P(X = x)$
0	0.029
1	0.049
2	0.078
3	0.155
4	0.212
5	0.262
6	0.215

SOLUTION You should still have a pair of columns named `x` and `Px` from Example 5.2; if not, create them first. Now enter the probability distribution shown in Table 5.3 into `x` and `Px` in the usual manner. Then proceed as follows.

Menu commands:

1 Choose Calc ➤ Calculator...
2 Type mu in the Store results in variable: text box
3 Select SUM from the Functions: drop down menu list
4 Enter `x` from the list box of column names
5 Click [*] on the calculator and enter `Px` from the list box of column names
6 Select [OK]

Session commands:

- Name an empty column `mu` if necessary
- Type `Let 'mu' = SUM(x * 'Px')` at the `MTB >` prompt and press [Enter←]

Now type `PRINT 'mu'` to see the result displayed as in Printout 5.3.

PRINTOUT 5.3 The mean of a random variable using Minitab

```
mu
  4.118
```

From Printout 5.3, the mean of the random variable X is stored in `'mu'` and hence is given by $\mu = 4.118$. ■

PROBLEM 5.3 Refer to Problem 5.1. In that problem you used Minitab to determine the probability distribution of the random variable Z—the number of persons in a randomly selected household. Now apply Minitab to the probability distribution you obtained to find the mean, μ, of the random variable Z. ◇

STANDARD DEVIATION OF A DISCRETE RANDOM VARIABLE

The *standard deviation*, σ, of a discrete random variable X is defined by

$$\sigma = \sqrt{\Sigma(x-\mu)^2 P(X=x)}$$

or, equivalently, by the computing formula

$$\sigma = \sqrt{\Sigma(x)^2 P(X=x) - \mu^2}$$

By using Minitab, we can alleviate the computations that must be performed to find the standard deviation of a random variable. Let us return to the situation of Example 5.3.

EXAMPLE 5.4 *Illustrates how to use Minitab to compute the standard deviation of a discrete random variable*

Use Minitab to find the standard deviation of the random variable X,. Recall that X is the number of tellers busy with customers at, say, 1:00 P.M. and has the probability distribution displayed in Table 5.3. Use the formula

$$\sigma = \sqrt{\Sigma(x)^2 P(X = x) - \mu^2}.$$

SOLUTION The probability distribution of the random variable X is shown in Table 5.3. We have already stored the x-values in x and the $P(X = x)$-values in Px. Also, the mean, μ, of the random variable X is stored in mu. Proceed as follows.

Menu commands:

1 Choose Calc ➤ Calculator...
2 Type sigma in the Store results in variable: text box
3 Press [Tab] and select Square root from the Functions: drop down menu list
4 Select Sum from the Functions: drop down menu list
5 Enter x from the list box of column names
6 Click the [**] on the calculator and type 2
7 Click [*] on the calculator and enter Px from the list box of column names
8 Press the right arrow key to move one space to the right
9 Click [-] on the calculator and enter mu from the list box of column names
10 Click [**] on the calculator and type 2
11 Select [OK]

Session commands:

- Name an empty column sigma if necessary
- Type Let 'sigma' = SQRT(SUM(x**2 * 'Px')-mu**2) at the MTB > prompt and press [Enter]

Now type PRINT 'sigma' to see the result displayed in Printout 5.4.

PRINTOUT 5.4 The standard deviation using Minitab

```
sigma
   1.57483
```

Thus the standard deviation of the random variable X is $\sigma = 1.57483$. ■

PROBLEM 5.4 Consider once more the random variable Z-in Problem 5.1. Use Minitab to determine the standard deviation, σ, of the random variable X using the computing formula of this section. ◇

A MACRO FOR THE MEAN AND STANDARD DEVIATION

Selecting the menu options or typing the foregoing commands in Minitab requires some care. For one thing, none of the commands may be misspelled. In addition, you must be sure to include parentheses just as they occur. This is a natural place to create a simple local macro to automate these computations and guarantee consistent computations.

In Lesson 1, you were asked to create a local macro called MUSIG.MAC listed in Printout 1.11 and to store that macro in your Macros folder. If you have not done so, please return to that lesson and carry out the exercise before proceeding. Let us apply the macro to the random variable in Example 5.3.

EXAMPLE 5.5 ***Illustrates the use of %MUSIG to compute the mean and standard deviation of a random variable***

Use the macro %MUSIG to find the mean and standard deviation of the random variable X, if X is the number of tellers busy with customers in Example 5.3.

SOLUTION Assume the data from Table 5.3 are located in columns named `x` and `Px`. If not you will have to enter the data in such columns, or in this instance, columns of your choice, named or unnamed. Printout 5.5 shows the result of executing this macro by typing `%MUSIG 'x' 'Px'` at the `MTB >` prompt and press Enter. (Please note the space between the arguments `'x'` and `'Px'`.)

PRINTOUT 5.5 Executing the macro %MUSIG

```
mu        4.11800
sigma     1.57483
```

You see that the answers agree completely with those we computed before. Again, that is not surprising since we merely mimicked the computation formulas. ■

PROBLEM 5.5 Use the macro %MUSIG to find the mean and standard deviation of the random variable Z, if Z is the random variable of Problem 5.1. ◇

5.3 The binomial distribution

E-5.6:I-5.3

Suppose that n Bernoulli trials are performed, with the probability of success on any given trial being p. Let X denote the total number of successes in the n trials. Then probabilities for the random variable X are given by the *binomial probability formula*

$$P(X = x) = \binom{n}{x} p^x (1-p)^{n-x}, \quad x = 0, 1, \ldots, n$$

The computations required for obtaining binomial probabilities using the binomial probability formula can be tedious and time-consuming. Fortunately, Minitab has two general commands, PDF and CDF, for computing various probabilities. Binomial probabilities are computed from these commands using a subcommand called BINOMIAL with a specification of the parameters n and p. Here is an example of how those programs can be used.

EXAMPLE 5.6 *Illustrates the BINOMIAL subcommand to PDF and CDF*

According to tables provided by the U.S. National Center for Health Statistics in *Vital Statistics of the United States*, there is about an 80% chance that a person age 20 will be alive at age 65. Suppose three people age 20 are selected at random. Find the probability that the number alive at age 65 will be

a) exactly two.
b) at most one.
c) at least one.
d) Determine the probability distribution for the number alive at age 65.
e) Construct a probability histogram for the number alive at age 65.

SOLUTION In this situation, a success is a person currently age 20 who will be alive at age 65. The probability of a success is thus $p = 0.8$. The number of trials is the number of people in the study so $n = 3$. Consequently, the appropriate binomial distribution is the one with parameters $n = 3$ and $p = 0.8$.

a) To obtain the probability that exactly two of the three people will be alive at age 65, two successes, we proceed as follows.

Menu commands:

1. Choose Calc ➤ Probability Distributions ➤ Binomial...
2. Select the Probability (the default) option button
3. Select the Number of trials: text box and type 3
4. Select the Probability of success: text box and type 0.8
5. Select the Input constant: option button, press [Tab] and type 2
6. Select [OK]

The resulting output is shown in Printout 5.6.

PRINTOUT 5.6
Minitab output for PDF 2; BINOMIAL 3, 0.8.

```
Binomial with n = 3 and p = 0.800000

        x        P( X = x)
     2.00          0.3840
```

The same results can be achieved with the session commands given on the following page.

Session commands:

- Type PDF 2; at the MTB > prompt and press [Enter]
- Type BINOMIAL 3 0.8. at the SUBC> prompt and press [Enter]

Therefore the probability is 0.3840 that exactly two of the three people will be alive at age 65. In the printout you are shown the parameters (n and p) that were used as well as the value of x and the result.

b) Here we want the probability that at most one person will be alive at the age of 65, that is, less than or equal to one. There are several ways to find that probability using Minitab. One way is to use the CDF command. The CDF program calculates *cumulative probabilities.* A *cumulative probability* gives the probability that a random variable will be *less than or equal to* (at most) a specified value. In this instance we want the cumulative probability (CDF) corresponding to the value 1.

Menu commands:

1 Choose Calc ➤ Probability Distributions ➤ Binomial...
2 Select the Cumulative probability option button
3 Enter the parameters as before and type 1 as the choice in the Input constant: text box
4 Select [OK]

Session commands:

- Type CDF 1; at the MTB > prompt and press [Enter]
- Type BINOMIAL 3, 0.8. at the SUBC> prompt and press [Enter]

See Printout 5.7 for result. Thus the probability of at most one person being alive at age 65 is 0.1040 (chances of about 10%).

PRINTOUT 5.7 Minitab output for CDF 1; BINOMIAL 3, 0.8.

```
Binomial with n = 3 and p = 0.800000

      x        P( X <= x)
   1.00          0.1040
```

c) The event at least one person will be alive at the age of 65 is the complement of the event no one will be alive at the age of 65. Hence the required probability is $1 - P(X = 0)$. Now $P(X = 0)$ may be computed exactly as in part (a) with 0 substituted for 2. The result is shown in Printout 5.8.

PRINTOUT 5.8 Minitab output for PDF 0; BINOMIAL 3, 0.8

```
Binomial with n = 3 and p = 0.800000

      x        P( X = x)
   0.00          0.0080
```

Consequently, the probability of at least one person being alive at the age of 65 is $1 - 0.008 = 0.992$.

d) Here we are to determine the probability distribution for the number alive at age 65. In other words, we want to obtain the binomial distribution with parameters $n = 3$ and $p = 0.8$. First, store the integers 0–3 in a column named x using patterned data entry as follows.

Menu commands:

1 Choose Calc ➤ Make Patterned Data ➤ Simple Set of Numbers...
2 Enter x in the Store patterned data in: text box
3 Select the From first value: text box and type 0
4 Select the To last value: text box and type 3
5 Select [OK]

Session commands:

- Name an empty column x if necessary
- Type SET 'x' at the MTB > prompt and press [Enter]
- Type 0:3 at the DATA> prompt and press [Enter]
- Type END at the next DATA> prompt and press [Enter]

Now we are ready to store the binomial probabilities. Proceed as follows.

Menu commands:

1 Choose Calc ➤ Probability Distributions ➤ Binomial...
2 Select the Probability option button
3 Enter the parameters 3 and 0.8 just as before
4 Select the Input column: option button, and enter x in the corresponding text box
5 Press [Tab] and type 'Px' in the Optional storage: text box
6 Select [OK]

Session commands:

- Name an empty column Px if necessary
- Type PDF 'x' 'Px'; at the MTB > prompt and press [Enter]
- Type BINOMIAL 3 0.8. at the SUBC> prompt and press [Enter]

Type Print 'x' 'Px' to see the results displayed in Printout 5.9 on the following page.

PRINTOUT 5.9
Binomial distribution with $n = 3$, $p = 0.8$

Row	x	Px
1	0	0.008
2	1	0.096
3	2	0.384
4	3	0.512

Step 5 in the menu commands may be omitted if there is no need to store probabilities. If you have no need to store either input or output, the same result may be accomplished with the simple session command as follows.

Session commands:

- Type PDF; at the MTB > prompt and press Enter
- Type Binomial 3 0.8. at the SUBC> prompt and press Enter

The resulting output is displayed in Printout 5.10.

PRINTOUT 5.10
Minitab output for PDF; BINOMIAL 3, 0.8.

```
Binomial with n = 3 and p = 0.800000

          x       P( X = x)
          0         0.0080
          1         0.0960
          2         0.3840
          3         0.5120
```

Compare these outputs with those found in *IS/6e* and *ES/5e*.

e) Now that we have the entire binomial probability distribution stored, we may use the Minitab GRAPH command to obtain a probability histogram of the distribution. Proceed as follows.

Menu commands:

1. Choose Graph > Chart...
2. Select the Y text box for Graph 1 and enter Px
3. Select the X text box for Graph 1 and enter x
4. Select the Edit Attributes... button, click the Bar Width button for Graph 1 and select 1.0; then select OK
5. Click the ▾ button for Frame and select Axis...
6. Select the Label text box for 1 and type Number of Heads; select the Label text box for 2 and type Probability
7. Select OK twice

The result may be seen in Figure 5.1 on the following page.

FIGURE 5.1 Minitab probability histogram for the Binomial 3, 0.8 distribution

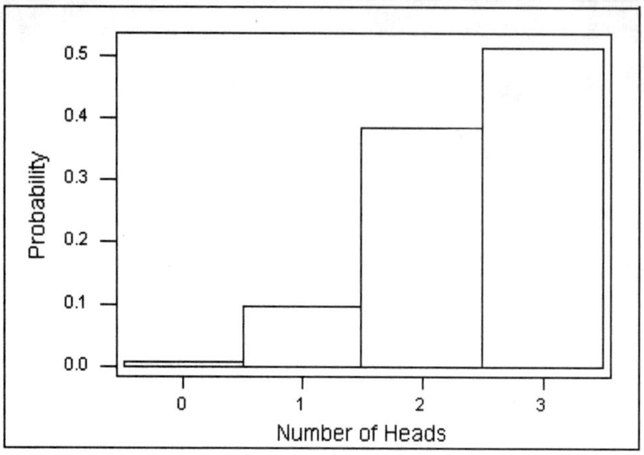

The same results may be obtained with the following session commands

Session commands:

- Type <u>Chart 'Px' * 'x';</u> at the MTB > prompt and press [Enter]
- Type <u>Bar;</u> at the SUBC> prompt and press [Enter]
- Type <u>Width 1.0;</u> at the SUBC> prompt and press [Enter]
- Type <u>ScFrame;</u> at the SUBC> prompt and press [Enter]
- Type <u>ScAnnotation;</u> at the SUBC> prompt and press [Enter]
- Type <u>Axis 1;</u> at the SUBC> prompt and press [Enter]
- Type <u>Label ''Number of Heads'';</u> at the SUBC> prompt and press [Enter]
- Type <u>Axis 2;</u> at the SUBC> prompt and press [Enter]
- Type <u>Label ''Probability''.</u> at the SUBC> prompt and press [Enter]

If your release does not support that much refinement, you may get just about the same information by using the following session command.

Session commands:

- Type <u>PLOT 'Px'*'x'</u> at the MTB > prompt and press [Enter]

This simple command produces a plot showing a single point above each x-value, not a histogram. Technically speaking, it is a truer representation of the probability distribution for a discrete random variable since the distribution consists of a set of isolated pairs of numbers. See Figure 5.2 on the following page for the Minitab output of this simple command.

Compare this figure with those found in *IS/6e* and *ES/5e*. From any of the displays, the left skewness is readily apparent. ■

FIGURE 5.2
Minitab plot for the Binomial 3, 0.8 distribution

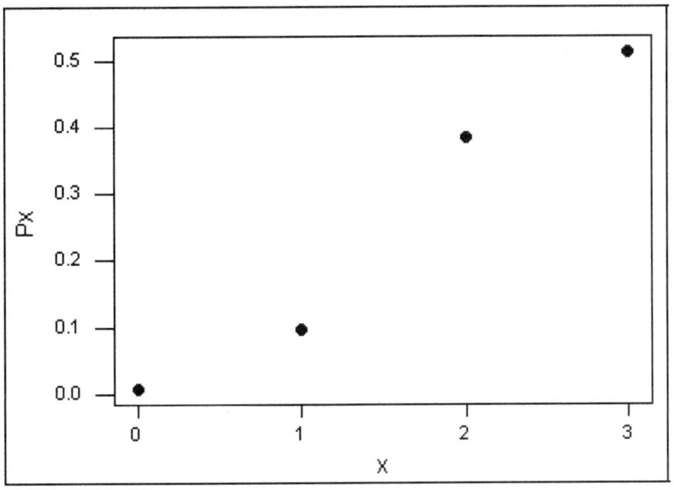

PROBLEM 5.6

According to a 1980 telephone study in Phoenix, the probability is about 0.25 that a randomly selected phone call will last longer than the mean duration of 3.8 minutes. What is the probability that, out of *three* randomly selected calls,

a) exactly two last longer than 3.8 minutes?
b) at most two last longer than 3.8 minutes?
c) at least one lasts longer than 3.8 minutes?
d) Determine the probability distribution for the number of calls out of 3 that last longer than 3.8 minutes.
e) Construct a graph for the number of calls, and identify the skewness of this binomial distribution. Use the commands given in Example 5.6 to carry out the computations. ◇

THE BINOMIAL MEAN AND STANDARD DEVIATION USING THE MACRO %MUSIG

Since the binomial distribution is just a special discrete distribution, we may employ the macro %MUSIG if we wish to compute the mean and standard deviation.

EXAMPLE 5.7 *Illustrates the use of %MUSIG for the binomial distribution*

According to a spokesperson for Southwest Airlines, the no-show rate for reservations is about 16%; that is, the probability is 0.16 that a party making a reservation will not show up. Use the macro MUSIG to find the mean and standard deviation of the number of no-shows for a flight having 42 parties with reservations.

First, store the integers 0–42 in `x` using patterned data entry and store the probabilities in a column named `Px` as in the last example. Actually, any two columns will do so long as you remember which contains the x-values and which contain the corresponding probabilities. Recall that these columns in which you store the data are passed as variables to the local variables c and p in the macro. With the column data stored, proceed to carry out the following simple session command.

Session commands:

- Type `%musig 'x' 'Px'` at the `MTB >` prompt and press Enter↵

Again, note the space between the arguments, `'x'` and `'Px'`. The results are displayed in Printout 5.11.

PRINTOUT 5.11 Using the macro %MUSIG on a binomial distribution

```
mu        6.72000
sigma     2.37588
```

You may easily verify these values by computing them using an ordinary calculator (or Minitab session commands of course) and Formula 5.2 of *IS/6e* (Formula 5.5 of *ES/5e*). ■

PROBLEM 5.7 Consider again the telephone study in Problem 5.6. Use the macro %MUSIG to compute the mean and standard deviation of that binomial distribution having parameters $n = 3$ and $p = 0.25$. ◇

5.4 The Poisson distribution

I-5.4

The Poisson distribution is regularly used to model the frequency with which a specified event occurs during a particular period of time, as well as a variety of other random phenomena.

The computations required for obtaining Poisson probabilities using the Poisson probability formula is complicated slightly by the ability to compute the factorial function $x!$ and the fact that there are, technically, infinitely many possible values of the random variable. Fortunately, Minitab has a POISSON subcommand that will work with the general commands PDF and CDF for computing various probabilities when the parameter λ (called mu in Minitab) is specified.† Here is an example of how those programs can be used.

EXAMPLE 5.8 Illustrates the POISSON subcommand to PDF and CDF

Desert Samaritan Hospital keeps detailed records of emergency-room traffic. From these records, we find the number of patients arriving between 6:00 PM and 7:00 PM

† As you will see in the next section, λ is the mean, μ, of X. You should know that in many other textbooks, λ is defined to be the reciprocal $1/\mu$.

has a Poisson distribution with parameter $\lambda = 6.9$. Use Minitab to compute the following probabilities:

a) exactly four patients arrive.
b) at most two patients arrive.
c) Determine the probability distribution for the number of patients arriving. Only use values for which the probability is positive to three decimal places.
d) Use part (c) to construct a (partial) probability histogram for the random variable X.

SOLUTION Let X denote the number of patients arriving between 6:00 PM and 7:00 PM. Then the random variable X has a Poisson distribution with parameter $\lambda = 6.9$.

a) To obtain the probability of exactly four patients arriving, we proceed as follows.

Menu commands:

1 Choose Calc ➤ Probability Distributions ➤ Poisson...
2 Select the Probability option button
3 Select the Mean: text box and type 6.9
4 Select the Input constant: option button, press Tab and type 4
5 Select OK

Session commands:

- Type `PDF 4;` at the `MTB >` prompt and press Enter
- Type `POISSON 6.9.` at the `SUBC>` prompt and press Enter

This produces the following Minitab output:

```
Poisson with mu = 6.90000

          x       P( X = x )
       4.00           0.0952
```

Therefore the probability is 0.0952 that exactly four patients will arrive during the time interval specified.

b) Here we want the probability that at most two patients arrive between 6:00 PM and 7:00 PM; that is, less than or equal to two arrivals. There are several ways to find that probability using Minitab. Here, we want the cumulative probability for the value 2. This suggests using the CDF command in Minitab. Proceed with the commands on the following page.

Menu commands:

1 Choose Calc ➤ Probability Distributions ➤ Poisson...
2 Select the Cumulative probability option button, enter the value of the Mean as before and type 2 as the choice in the Input constant: text box
3 Select OK

Session commands:

- Type CDF 2; at the MTB > prompt and press Enter
- Type POISSON 6.9. at the SUBC> prompt and press Enter

Either command produces the following Minitab output:

```
Poisson with mu = 6.90000

        x     P( X <= x )
     2.00          0.0320
```

Consequently, the probability of at most two arrivals is 0.0320.

c) Next, we are to determine the probability distribution for the number of arrivals. In other words, we want to obtain the Poisson distribution with parameter $\lambda = 6.9$. The easiest way to accomplish this is with the following simple session commands.

Session commands:

- Type PDF; at the MTB > prompt and press Enter
- Type POISSON 6.9. at the SUBC> prompt and press Enter

Printout 5.12 on the following page shows the output that results from this session command. Notice that probabilities are so small starting with $x = 20$ that no values beyond this are listed even though, technically, there is a positive probability associated with each value of X, no matter how large. In order to use menu commands you would have to know ahead of time at what point probabilities are negligible or else enter a large number of values in x and proceed as we did in the binomial case. Compare this table with Table 5.16 in *IS/6e* .

d) In order to use the GRAPH or PLOT command, we need to have these values stored. As in the binomial case, we can accomplish this as follows. First store the integers 0–20 in x using patterned data entry as before. Now we can compute and store the probabilities using these commands.

Menu commands:

1 Choose Calc ➤ Probability Distributions ➤ Poisson...
2 Select the Probability option button
3 Enter the mean 6.9 as before
4 Select the Input column: option button, and enter x in the corresponding text box
5 Press Tab and type Px in the Optional storage: text box
6 Select OK

The same results may be achieved by carrying out the session commands that are listed after the printout.

PRINTOUT 5.12
Minitab output for PDF; POISSON 6.9.

```
Poisson with mu = 6.90000

        x          P( X = x)
        0            0.0010
        1            0.0070
        2            0.0240
        3            0.0552
        4            0.0952
        6            0.1511
        7            0.1489
        8            0.1284
        9            0.0985
       10            0.0679
       11            0.0426
       12            0.0245
       13            0.0130
       14            0.0064
       15            0.0029
       16            0.0013
       17            0.0005
       18            0.0002
       19            0.0001
       20            0.0000
```

Session commands:

- Name an empty column `P(X=x)` if necessary
- Type <u>`PDF 'P(X=x)';`</u> at the `MTB` > prompt and press [Enter↵]
- Type <u>`POISSON 6.9.`</u> at the `SUBC>` prompt and press [Enter↵]

Now that the probability distribution is stored in the Minitab worksheet, we may use the GRAPH command to examine the behavior of this particular Poisson distribution. Proceed as follows.

Menu commands:

1 Choose Graph > Chart...
2 Select the Y text box for Graph 1 and enter Px
3 Select the X text box for Graph 1 and enter x
4 Select the Edit Attributes... button, click the Bar Width button for Graph 1 and select <u>1.0</u>; then click [OK]
5 Click the [▾] button for Frame and select Axis...
6 Select the Label text box for 1 and type <u>`Number of Patients`</u>; select the Label text box for 2 and type <u>`Probability`</u>
7 Click [OK] twice

These steps produce the following histogram shown on the following page.

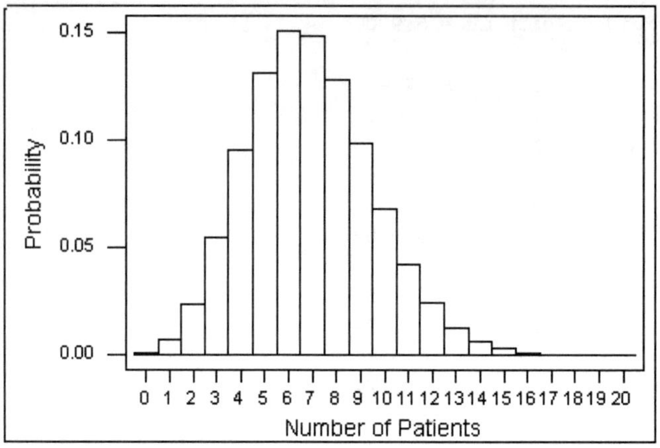

The same results may be obtained with the following session commands.

Session commands:

- Type Chart 'Px' * 'x'; at the MTB > prompt and press Enter
- Type Bar; at the SUBC> prompt and press Enter
- Type Width 1.0; at the SUBC> prompt and press Enter
- Type ScFrame; at the SUBC> prompt and press Enter
- Type ScAnnotation; at the SUBC> prompt and press Enter
- Type Axis 1; at the SUBC> prompt and press Enter
- Type Label ''Number of Patients''; at the SUBC> prompt and press Enter
- Type Axis 2; at the SUBC> prompt and press Enter
- Type Label ''Probability''. at the SUBC> prompt and press Enter

If your release does not support these commands, you may produce the following plot by typing PLOT 'Px'*'x' at the MTB > prompt.

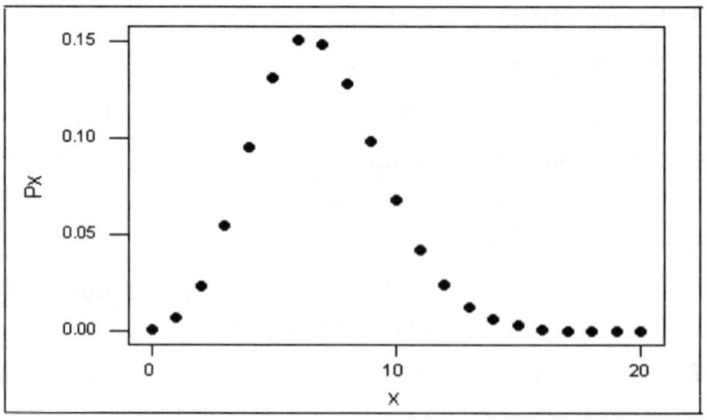

This simple plot conveys about the same amount of information as the probability histogram. Indeed, since the probabilities are single points at the corresponding x-values, the plot gives a more accurate representation of the appearance of the distribution.

As you can see from either figure, this particular distribution is skewed slightly to the right. Most of the probability is associated with relatively small values of `x` with a trailing off to the right. On the other hand, there is a great deal of symmetry in the range of x-values 0–16, where most of the probability is located. Indeed, $P(x > 16) = 0.0008$. You may verify this in Minitab using the CDF command with the POISSON subcommand as in part (b) in this example. You would discover directly that $P(x \leq 16) = 0.9992$ if you type `CDF 16;` followed by `POISSON 6.9.` as a subcommand. ■

PROBLEM 5.8 Assume that the sales made in a week by a used car salesperson occur like a Poisson random variable with parameter $\lambda = 1$. What is the probability that the salesperson makes,

a) exactly two sales in a week?
b) at most two sales in a week?
c) Determine the probability distribution for the number sales in a week.
d) Use the results to plot the distribution and discuss its shape. ◇

THE POISSON MEAN AND STANDARD DEVIATION USING THE MACRO %MUSIG

Since the Poisson distribution is just a special discrete distribution, we may employ the macro %MUSIG if we wish to compute the mean and standard deviation. In this way, you may use the computer to obtain more verification of values given in Formula 5.4 in *IS/6e* . Of course we already know that the mean is 6.9 since we have been using that value for input in the functions.

EXAMPLE 5.9 *Illustrates the use of %MUSIG for the Poisson distribution*

Use the macro %MUSIG to find the mean and standard deviation for the Poisson distribution having parameter $\lambda = 6.9$. Verify that these are the results you obtain from Formula 5.4 in *IS/6e* . We assume you still have columns named `x` and `Px`, where the distribution is stored.

SOLUTION To process the macro, just type `%MUSIG 'x' 'Px'`. (Please note the space between the arguments.) The results are displayed in Printout 5.13 and must be approximations since the entire distribution is not stored.

PRINTOUT 5.13
Using the macro %MUSIG on a Poisson distribution

```
mu       6.89975
sigma    2.62642
```

From Printout 5.13, you see that the mean number patient arrivals is 6.89975 (technically equal to the parameter 6.9) with a standard deviation of 2.62642. ■

PROBLEM 5.9 Consider again the Poisson assumptions in Problem 5.8. Use the macro %MUSIG to compute approximate values of the mean and standard deviation of the Poisson distribution having parameters $\lambda = 1$. Compare the results with Formula 5.4 of *IS/6e* . ◇

POISSON APPROXIMATION TO THE BINOMIAL DISTRIBUTION

When n is large and p is small, the Poisson formula with $\lambda = np$ for probabilities can be used to approximate the binomial probabilities having parameters n and p. A general rule of thumb in order to apply the approximation are the two conditions $n \geq 100$ and $np \leq 10$. When computations had to made by hand, this was a very valuable computation tool. With the computational capabilities we now have at our disposal, the approximation is much less valuable. Many computing routines, including the BINOMIAL command in Minitab, can handle a wide range of parameter choices, making the Poisson approximation unnecessary. Let us illustrate.

EXAMPLE 5.10 *Illustrates the Poisson approximation*

According to the *World Almanac*, the infant mortality rate in Sweden is 4.5 per 1000 live births. Determine the probability that, out of 500 randomly selected live births, there are

a) no infant deaths.
b) at most three infant deaths. Use Minitab and the Poisson approximation to compute these probabilities.
c) Compare the results with the corresponding binomial probabilities using Minitab's BINOMIAL command.

SOLUTION Clearly the problem as stated is a binomial model with $n = 500$ and $p = 0.0045$. This is a case where the Poisson approximation applies, with $np = 2.25$.

a) To obtain the probability of no infant deaths, we proceed as follows.

Menu commands:

1 Choose Calc ➤ Probability Distributions ➤ Poisson...
2 Select the Probability (the default) option button
3 Select the Mean: text box and type 2.25
4 Select the Input constant: option button, press [Tab] and type 0
5 Select [OK]

This produces the following Minitab output:

```
Poisson with mu = 2.25000

          x       P( X = x)
       0.00          0.1054
```

The same result may be obtained using the following session commands.

Session commands:

- Type PDF 0; at the MTB > prompt and press Enter
- Type POISSON 2.25. at the SUBC> prompt and press Enter

Therefore the probability is 0.1054 that there are no infant deaths.

b) Here we need to use the CDF command.

Menu commands:

1. Choose Calc ➤ Probability Distributions ➤ Poisson...
2. Select the Cumulative probability option button
3. Select the Mean: text box and type 2.25
4. Select the Input constant: option button, press Tab and type 3
5. Select OK

Session commands:

- Type CDF 3; at the MTB > prompt and press Enter
- Type POISSON 2.25. at the SUBC> prompt and press Enter

This produces the following Minitab output:

```
Poisson with mu = 2.25000

          x      P( X <= x)
       3.00          0.8094
```

Consequently, the probability of at most three infant deaths is 0.8094.

c) Finally, we are to determine these probabilities using the BINOMIAL command. The procedure is straightforward.

Menu commands:

1. Choose Calc ➤ Probability Distributions ➤ Binomial...
2. Select the Probability (the default) option button
3. Select the Number of trials: text box and type 500
4. Select the Probability of success: text box and type 0.0045
5. Select the Input constant: option button, press Tab , and type 0

6 Select OK

Repeat the foregoing menu commands selecting the Cumulative probability option button at Step 2 and typing 3 as the choice in the Input constant: text box at Step 5.

Session commands:

- Type PDF 0; at the MTB > prompt and press Enter
- Type BINOMIAL 500, 0.0045. at the SUBC> prompt and press Enter

Repeat the foregoing menu commands typing CDF 3; in place of PDF 0;. This produces the following Minitab output:

```
Binomial with n = 500 and p = 0.00450000

        x          P( X = x)
     0.00             0.1049

        x          P( X <= x)
     3.00             0.8098
```

Notice that these more exact probabilities are approximated satisfactorily by the Poisson approximate values. However, it was a simple matter to compute the more exact values, making the approximating values redundant. Thanks to modern technology and the extent to which our more exact routines function, approximations are becoming less and less important. ■

PROBLEM 5.10 Using Example 5.10 as a model, compute the probabilities that there will be
a) exactly two infant deaths,
b) at most four infant deaths,
using both the Poisson approximation formulas and the more exact Binomial formulas. ◇

LESSON 6

The Normal Distribution

GENERAL OBJECTIVE

Of all the probability distributions used in the study of both probability and statistics, the *normal distribution* is the most important. It is often appropriate to use a normal distribution as the distribution of a population or random variable. Furthermore, the normal distribution is applied extensively in inferential statistics. In this lesson we will learn how Minitab can be employed to perform a variety of tasks related to the normal distribution.

LESSON OUTLINE

6.1 Normally distributed random variables

E-6.1:I-6.1

We may use Minitab's RANDOM command to advantage to simulate a normally distributed random variable. We can then plot a histogram of the resulting data and compare the result to a normal curve having the same parameters, namely mean and standard deviation, as the following example shows.

EXAMPLE 6.1 *Illustrates the simulation of a normal distribution*

Gestation periods of humans, X, are normally distributed with a mean of 266 days and a standard deviation of 16 days. Use Minitab to simulate 1000 human gestation periods and then obtain a histogram of the results, comparing the histogram to a normal curve having $\mu = 266$ and $\sigma = 16$.

SOLUTION We use the RANDOM command in Minitab to simulate the actual performance of the experiment involving observing the value of X. That is, we imitate the experiment consisting of 1,000 independent trials or observations from this normal distribution. Proceed as follows.

Menu commands:

1 Choose Calc ➤ Random Data ➤ Normal...
2 Type 1000 in the Generate [ ] rows of data text box
3 Select the Store in column(s): text box and type DAYS
4 Select the Mean: text box and type 266
5 Select the Standard deviation: text box and type 16
6 Select [OK]

Session commands:

- Name an empty column DAYS
- Type `RANDOM 1000 'DAYS';` at the MTB > prompt and press [Enter↵]
- Type `NORMAL 266, 16.` at the SUBC> prompt and press [Enter↵]

This much places 1,000 observations from the normal distribution in the column DAYS. To create a histogram of these data and superimpose a normal curve having the same parameters, proceed as follows.

Menu commands:

1 Choose Stat ➤ Basic Statistics ➤ Display Descriptive Statistics...
2 Enter DAYS in the Variables: text box
3 Select the Graphs... button
4 Select the Histogram of data, with normal curve check box
5 Select [OK] twice

Session commands:

- Type HISTOGRAM 'DAYS' at the MTB > prompt and press Enter (There is no normal curve option.)

The result is the histogram found in Figure 6.1. In addition, the menu command Graphs option selected in Step 4 superimposes a normal curve having $\mu = 266$ and $\sigma = 16$ for comparison.

FIGURE 6.1
Histogram of 1000 simulated gestation periods

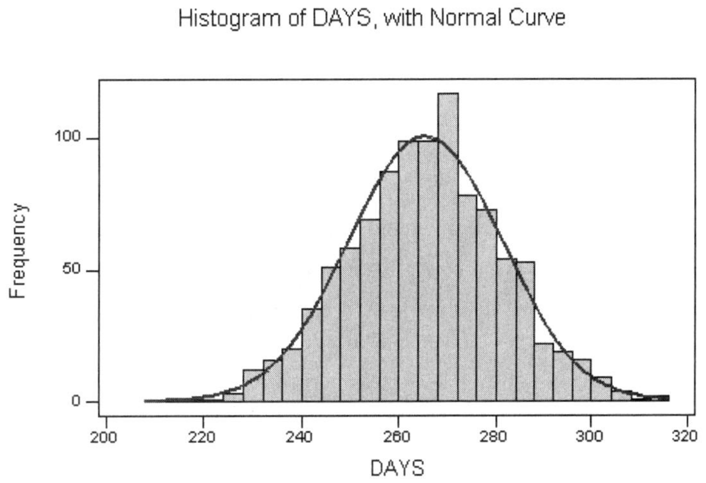

As expected, the histogram is roughly bell-shaped and compares favorably with the superimposed theoretical distribution. Naturally, your histogram will not be identical to this one due to the nature of simulation. ■

PROBLEM 6.1 Using the circumstances of Example 6.1, simulate 1000 gestation periods and use Minitab to construct a histogram of the results. ◇

6.2 Areas under the standard normal curve **E-6.2:I-6.2**

Probabilities for normally distributed populations or random variables are equal to areas under a suitable "bell-shaped" curve. Such a curve is called a *normal curve.* There are many, in fact, infinitely many normal curves. In this section we will consider a special normal curve—the *standard normal curve* or *z-curve.*

Areas under the standard normal curve can be determined with the aid of standard normal tables like Table II found in *IS/6e* and *ES/5e*. Alternatively, we can use Minitab to find areas under the standard normal curve. The appropriate

Minitab command is CDF (without a subcommand in a session window). This command computes the area under the standard normal curve that lies to the *left* of a specified value of z, that is, a z-score.

EXAMPLE 6.2 Illustrates the Normal CDF

Use Minitab to find the area under the standard normal curve that lies
a) to the left of $z = 1.23$.
b) to the right of $z = 0.76$.
c) between $z = -0.68$ and $z = 1.82$.

SOLUTION a) As we pointed out earlier, the command CDF computes the area under the standard normal curve to the left of a z-score.

Menu commands:

1 Choose Calc ➤ Probability Distributions ➤ Normal...
2 Select the Cumulative probability radio button
3 Select the Mean: text box and type 0; select the Standard deviation: text box and type 1 (if necessary)
4 Select the Input constant: radio button, press [Tab] and type 1.23 in the text box
5 Select [OK]

Session commands:

- Type CDF 1.23 at the MTB > prompt and press [Enter]

This produces the following Minitab output:

```
Normal with mean = 0 and standard deviation = 1.00000

          x      P( X <= x )
     1.2300          0.8907
```

The area under the standard normal curve to the left of $z = 1.23$ is 0.8907.

b) Here we want to find the area under the standard normal curve to the right of $z = 0.76$. Since CDF computes areas to the *left* of specified z-scores, we proceed as follows.

Menu commands:

1 Choose Calc ➤ Probability Distributions ➤ Normal...
2 Select the Cumulative probability radio button
3 Select the Mean: text box and type 0; select the Standard deviation: text box and type 1

4 Select the Input constant: radio button, press Tab and type 0.76 in the text box

5 Select the Optional storage: text box and type K1 (or any other constant of your choice)

6 Select OK

Session commands:

- Type CDF 0.76 K1 at the MTB > prompt and press Enter

This stores the area to the left of $z = 0.76$ in K1. Because the total area under the standard normal curve is 1, the area to the right of $z = 0.76$ equals 1 minus the area to the left of $z = 0.76$. Hence we can now get the area to the right of $z = 0.76$ by typing LET K2=1-K1 followed by PRINT K2. This produces the following Minitab output:

```
MTB > LET K2=1-K1
MTB > PRINT K2
 K2     0.223627
```

c) The area under the standard normal curve between $z = -0.68$ and $z = 1.82$ equals the area to the left of $z = 1.82$ *minus* the area to the left of $z = -0.68$. We can use CDF to obtain these latter two areas simultaneously. To do so, first store the two z-scores, $z = -0.68$ and $z = 1.82$, into a column named VAL in the usual way. We can then have the CDF command find the areas to the left of each value in the column and store the results in another column as follows.

Menu commands:

1 Choose Calc ➤ Probability Distributions ➤ Normal...

2 Select the Cumulative probability radio button

3 Select the Mean: text box and type 0; select the Standard deviation: text box and type 1

4 Select the Input column: radio button

5 Press Tab and enter VAL for the column

6 Select the Optional storage: text box and type PVAL

7 Select OK

Session commands:

- Name an empty column PVAL if necessary
- Type CDF 'VAL' 'PVAL' at the MTB > prompt and press Enter

Now the areas under the standard normal curve to the left of $z = -0.68$ and $z = 1.82$ are stored, respectively, in rows 1 and 2 of PVAL. To determine the area between $z = -0.68$ and $z = 1.82$, type LET K1='PVAL'(2)-'PVAL'(1) and PRINT K1. This produces the following Minitab output:

```
MTB > LET K1='PVAL'(2)-'PVAL'(1)
MTB > PRINT K1
K1    0.717368
```

Consequently, we see that the area under the standard normal curve between $z = -0.68$ and $z = 1.82$ equals 0.717368. The area under the standard normal curve between any two z-scores can be determined in the same way. ■

PROBLEM 6.2 Use Minitab to find the area under the standard normal curve that lies

a) to the left of $z = -1.28$.
b) between $z = 1.53$ and $z = 2.47$.
c) to the right of $z = 1.64$.
d) either to the left of $z = -1.85$ or to the right of $z = 0.29$. ◇

FINDING THE z-SCORE FOR A SPECIFIED AREA

Minitab also has a command, called INVCDF, with the same subcommands as CDF. When executed without a subcommand, it will find the z-score for any specified area. The name INVCDF stands for the *inverse* of the CDF. It corresponds to going into the body of Table II of *IS/6e* and *ES/5e* and finding the value of z—the inverse of finding the area for a given value of z. Since the CDF command yields areas all the way to the left of z, the command INVCDF requires that you enter such areas. Please continue your reading in *IS/6e* and *ES/5e* through Definition 6.3 before proceeding with this lesson so that you will be acquainted with the z_α notation.

EXAMPLE 6.3 *Illustrates how to find the z-score for a specified area using Minitab*

Use Minitab to find the z-score for which the area under the standard normal curve to the left of that value is 0.04.

SOLUTION Since the area to the left is specified, this is the proper entry for the INVCDF command.

Menu commands:

1 Choose Calc ➤ Probability Distributions ➤ Normal...

2 Select the Inverse cumulative probability radio button.

3 Select the Mean: text box and type 0; select the Standard deviation: text box and type 1

4 Select the Input constant: radio button

5 Press Tab and type 0.04

6 Select OK

Session commands:

- Type INVCDF 0.04 at the MTB > prompt and press Enter

Printout 6.1 shows the result. If you used the menu procedure, you will also view `Normal 0.0 1.0` as a subcommand.

PRINTOUT 6.1
Output of INVCDF

```
Normal with mean = 0 and standard deviation = 1.00000

 P( X <= x)          x
    0.0400       -1.7507
```

The answer is -1.7507. This would be denoted $z_{0.96}$ in the standard notation and $z_{0.96} = -z_{0.04}$ from the symmetry property of the normal curve. ■

Note that, for all choices of $\alpha > 0.5$, the value of z_α will be negative.

EXAMPLE 6.4 Illustrates how to find the z-score for a specified area using Minitab

Use Minitab to find the z-value for which the area under the standard normal curve to the right of that value is 0.025.

SOLUTION To find the value using Minitab, we must input the area to the *left*, which is 0.975. Proceed as follows.

Menu commands:

1 Choose Calc ➤ Probability Distributions ➤ Normal...
2 Select the Inverse cumulative probability radio button
3 Select the Mean: text box and type 0; select the Standard deviation: text box and type 1
4 Select the Input constant: radio button
5 Press Tab and type 0.975
6 Select OK

Session commands:

- Type INVCDF 0.975 at the MTB > prompt and press Enter

Printout 6.2 on the following page shows the result.

PRINTOUT 6.2
Output of INVCDF

```
Normal with mean = 0 and standard deviation = 1.00000

 P( X <= x)          x
    0.9750      1.9600
```

As you see in Printout 6.2, Minitab reminds you of the input area and then prints the value 1.9600. You recognize this as $z_{0.025}$ as defined in Definition 6.3 in *IS/6e* and *ES/5e*. ■

In this way you may find the value of z_α for any choice of α. But you must remember to input $1-\alpha$ to the INVCDF command. One of the most common mistakes made in finding z_α, with or without the computer, is to use α for the area to the left of the point.

EXAMPLE 6.5 *Illustrates how to find the z-scores for a specified area using Minitab*

Use Minitab to determine the two z-scores that divide the area under the standard normal curve into a middle 0.95 area and two outside 0.025 areas. See Figure 6.2.

FIGURE 6.2
The standard normal curve

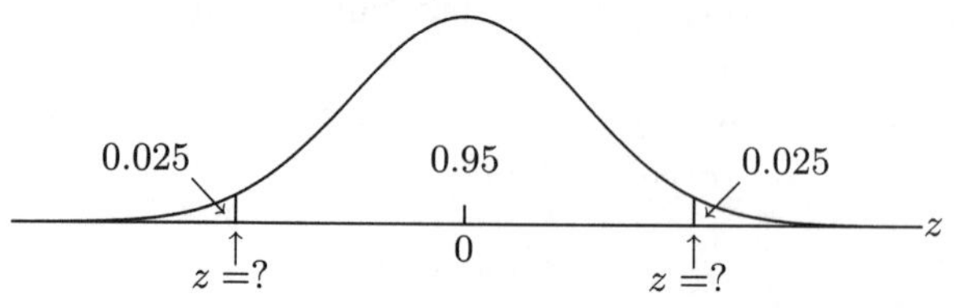

SOLUTION As we see from Figure 6.2, the area of the small region on the right-hand side is 0.025. This means that the z-score on the right is $z_{0.025}$. In Example 6.3, we found that $z_{0.025} = 1.96$. Thus the z-score on the right is 1.96. Since the standard normal curve is symmetric about 0, it follows that the z-score on the left is -1.96. ■

PROBLEM 6.3

For the standard normal curve

a) find the z-value with area 0.75 to its left.
b) find the z-value with area 0.95 to its right.
c) find the two z-values that divide the area into a middle 0.90.
d) find $z_{0.06}$ and $z_{0.94}$. ◇

6.3 Working with normally distributed random variables

E-6.3:I-6.3

The Minitab commands CDF and INVCDF, will accept arbitrary parameters μ and σ with the NORMAL subcommand. These can be used to compute areas to the *left* of a specified value as well as return x-values for any specified area to the left of x.

AREAS UNDER ANY NORMAL CURVE

In Procedure 6.1 of *IS/6e* and *ES/5e* you learned how to find a percentage or probability for any normally distributed variable. Minitab makes it unnecessary to calculate the intermediate z-scores to find such areas, Step 3 of that procedure, since any mean μ and standard deviation σ may be used in the Minitab command. Here is an example.

EXAMPLE 6.6 *Illustrates finding a probability for a normally distributed variable*

Intelligence quotients (IQ's) measured on the Stanford Revision of the Binet-Simon Intelligence Scale are known to be normally distributed with a mean of 100 and a standard deviation of 16. Obtain the percentage of people having IQs between 115 and 140.

SOLUTION Recall that CDF computes areas to the *left* of a specified value. We want to find the area under the normal curve with parameters $\mu = 100$ and $\sigma = 16$ that lies between 115 and 140. To do so, first store the two x-values, $x = 115$ and $x = 140$, into the column named VAL in the usual way. As in the case of z-values, we are going to find areas to the left of each of these values and store them in the column named PVAL. The difference between column entries will give the answer. Proceed as follows.

Menu commands:

1 Choose Calc ➤ Probability Distributions ➤ Normal...
2 Select the Cumulative probability radio button
3 Select the Mean: text box and type 100
4 Select the Standard deviation: text box and type 16
5 Select the Input column: radio button
6 Press [Tab] and enter VAL for the column
7 Select the Optional storage: text box and type PVAL
8 Select [OK]

Session commands:

- Name an empty column PVAL (if necessary)
- Type CDF 'VAL' 'PVAL' at the MTB > prompt and press [Enter]

Now the areas under the normal curve to the left of $x = 115$ and $x = 140$ are stored, respectively, in rows 1 and 2 of PVAL. To determine the percentage or scores between $x = 115$ and $x = 140$, execute the following session commands.

Session commands:

- Type LET K1='PVAL'(2)-'PVAL'(1)
- Type PRINT K1

Printout 6.3 displays the session commands and the result.

PRINTOUT 6.3 Area between two *x*-values

```
MTB > LET K1='PVAL'(2)-'PVAL'(1)
MTB > PRINT K1
K1        0.168041
```

Consequently, we see that approximately 16.8% of the population have IQs between 115 and 140. ■

PROBLEM 6.4 Use Minitab to find the area under the normal curve with parameters $\mu = 15$ and $\sigma = 2$ that lies

a) to the left of 12.
b) between 14 and 18.
c) to the right of 16.5. ◇

FINDING THE x-VALUE FOR A SPECIFIED PERCENTAGE

Recall that Minitab also allows the subcommand NORMAL, with specification of arbitrary parameters to the INVCDF command. That means you can find the x-value corresponding to any area to the left of that point, just as you were able to find the z-score in the standard normal curve.

EXAMPLE 6.7 *Illustrates finding the x-value for a specified percentage using INVCDF*

For the circumstances of Example 6.6, find the 90th percentile for IQs

SOLUTION We need to find the x-value corresponding to an area of 0.90 to the left of that value. We have only to input this value to the INVCDF command as follows.

Menu commands:

1 Choose Calc ➤ Probability Distributions ➤ Normal...
2 Select the Inverse cumulative probability radio button
3 Select the Mean: text box and type 100
4 Select the Standard deviation: text box and type 16

5 Select Input constant: radio button, press [Tab] and type 0.9 in the text box.

6 Select [OK]

Session commands:

- Type INVCDF 0.9; at the MTB > prompt and press [Enter]
- Type NORMAL 100, 16. at the SUBC> prompt and press [Enter]

Printout 6.4 shows the result.

PRINTOUT 6.4
The x-value for a specified area of 0.90

```
Normal with mean = 100.000 and standard deviation = 16.0000

 P( X <= x)          x
    0.9000       120.5048
```

From the printout we see that the 90th percentile is about 120.5. Thus, approximately 90% of the population has an IQ less that 120.5. ■

PROBLEM 6.5 For the normal curve with parameters $\mu = 64.6$ and $\sigma = 2.4$, use Minitab to find the x-value with
a) area 0.90 to its left.
b) area 0.90 to its right. ◇

6.4 Assessing normality; normal probability plots

E-6.4:I-6.4

Normal probability plots are useful tools for indicating whether it is reasonable to assume that the population under discussion can be represented by a normal distribution. Minitab has a command called NSCORES that will convert a set of data to normal scores for you. Then we may use the PLOT command to produce a normal probability plot. We illustrate this with the next example.

EXAMPLE 6.8 Illustrates the NSCORES command

In Table 6.1, we repeat the data for 12 IRS returns, as discussed in *IS/6e* and *ES/5e*.

TABLE 6.1
Adjusted gross incomes ($1000)

9.7	93.1	33.0	21.2
81.4	51.1	43.5	10.6
12.8	7.8	18.1	12.7

Use Minitab to construct a probability plot for the data (WeissStats CD:Tb06-03).

SOLUTION First, store the data in a column named AGI. Then proceed as follows.

Menu commands:

1 Choose Calc ➢ Calculator...
2 Type NSCORE in the Store result in variables: text box
3 Select Normal scores from the Functions menu choices
4 Replace number by AGI in the Expression: text box
5 Select OK

Session commands:

- Name an empty column NSCORE
- Type NSCORES 'AGI' 'NSCORE' at the MTB > prompt and press Enter↵

This much converts the incomes to normal scores, and you may see this conversion if you wish by just typing PRINT 'AGI''NSCORE'. Our primary interest, however, is in the plot of NSCORE versus AGI. We can accomplish this as follows.

Menu commands:

1 Choose Graph ➢ Plot...
2 Enter NSCORE in the Y text box for Graph 1 in the Graph variables: text box
3 Enter AGI in the X text box for Graph 1 in the Graph variables: text box
4 Select OK

Session commands:

- Type PLOT 'NSCORE'*'AGI' at the MTB > prompt and press Enter↵

The results are shown in the Figure 6.3.

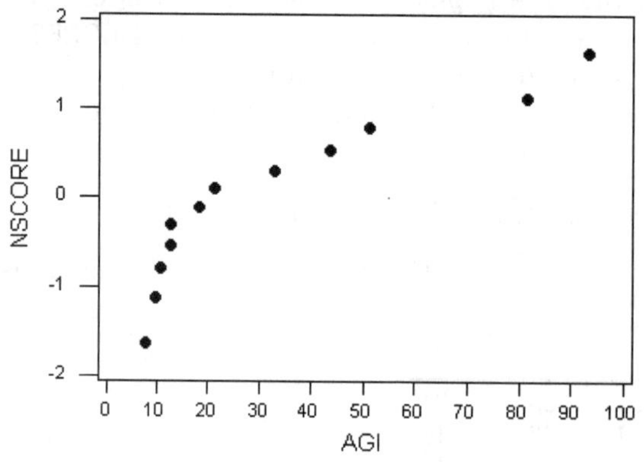

FIGURE 6.3 Minitab normal probability plot for adjusted gross incomes

You will notice considerable curvature in the probability plot indicating that AGI values are not normally distributed. ■

PROBLEM 6.6 The exam scores for the students in an introductory statistics course are as follows.

88	67	64	76	86
85	82	39	75	34
90	63	89	90	84
81	96	100	70	96

Store the data in **EXAM**. Use Minitab to construct a normal probability plot for these data, and make a judgment about assuming normality for these data. ◇

DETECTING OUTLIERS

Normal probability plots can help us make judgments about the nature of outliers in a data set, that is, items that fall well outside the data base.

EXAMPLE 6.9 Using Normal Probability Plots to Detect Outliers

The data on chicken consumption in the United States reported in *IS/6e* and *ES/5e* are repeated here in Table 6.2 (WeissStats CD:Tb06-05).

TABLE 6.2 Sample of last year's chicken consumption (lbs)

47	39	62	49	50	70
59	53	55	0	65	63
53	51	50	72	45	

Use Minitab to detect outliers in the data.

SOLUTION A probability plot for the data is shown in Figure 6.4.

FIGURE 6.4 Minitab normal probability plot for chicken consumption

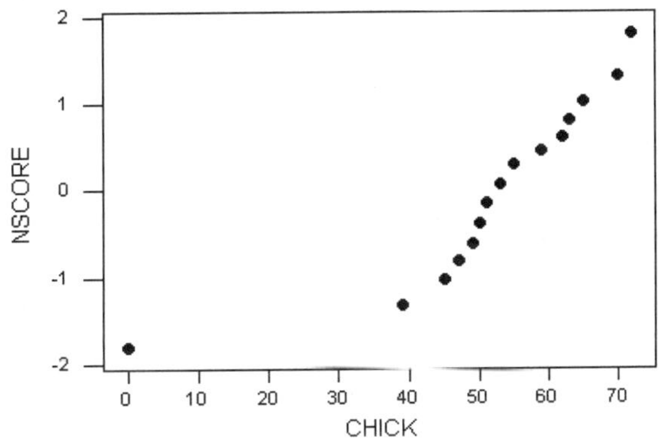

We obtained the normal probability plot in Figure 6.4 after storing the data in a column named CHICK. We continue to use the column NSCORE for storing the results of applying NSCORES to different columns of data. If you wish to preserve the data already there, then you may choose a different column designation, such as NSCORE2 for new data.

Notice that the data fall roughly in a straight line except for the value at 0. This would suggest that the subpopulation of people who consume no chicken at all are not representative of the population of chicken consumers. In other words, 0 appears to be an outlier so we remove it from the sample. When that is done, the probability plot then appears as in Figure 6.5.

FIGURE 6.5 Minitab normal probability plot for adjusted chicken consumption

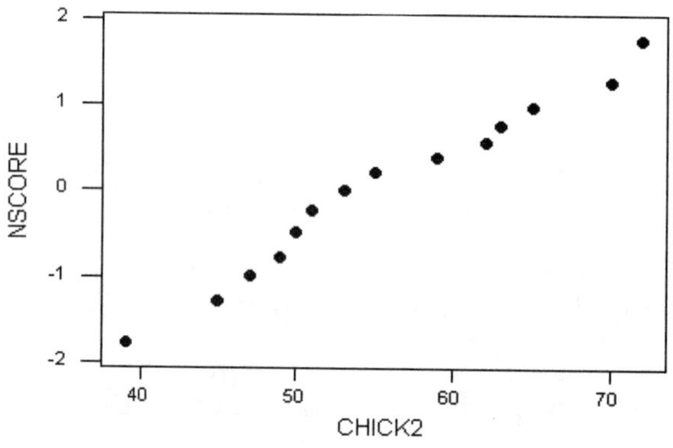

The new sample is more reasonably linear. Hence, the assumption of normality appears to be safe to use for these data. Perhaps it would be appropriate also to adjust our conclusions to apply only to chicken consumers in the population as a whole rather than the general population. ■

PROBLEM 6.7

The U.S. Federal Highway Administration conducts studies on motor vehicle travel by type of vehicle. Results are published annually in *Highway Statistics*. A sample of 15 cars yields the following data on number of miles driven, in thousands, for last year.

10.2	10.3	8.9	12.7	8.3
9.2	13.7	7.7	3.3	10.6
11.8	6.6	8.6	5.7	12.0

Store the data in MILES, and use Minitab to analyze these data for outliers. ◇

LESSON 7

The Sampling Distribution of the Mean

GENERAL OBJECTIVE

Minitab is also quite useful for amplifying and clarifying concepts related to sampling and sampling distributions. In this lesson we will employ Minitab to sample from a population to illustrate sampling error, the central limit theorem, and the sampling distribution of the mean.

LESSON OUTLINE

7.1 Sampling Error

E-7.1:I-7.1

A sample from a population provides us with only a portion of the entire population. Thus it is unreasonable to expect that the sample will yield perfectly accurate information about the population. Consequently, we should anticipate that a certain amount of error will result simply because we are sampling. This type of error is aptly called *sampling error.* We illustrate with an example.

EXAMPLE 7.1 *Illustrates sampling error*

Suppose that the *population* under consideration consists of the annual salaries of the top five state officials of Oklahoma. This population of salaries is presented in the second column of Table 7.1. [Data from *The World Almanac, 1985.*] Values are in thousands of dollars.

TABLE 7.1
Top five Oklahoma state officials and their annual salaries

Official	Salary ($1000s)
Governor (G)	70
Lieutenant Governor (L)	40
Secretary of State (S)	37
Attorney General (A)	55
Treasurer (T)	50

Use Minitab to draw a sample of size 2 and compare the sample mean with the population mean.

SOLUTION First, store the population of five salaries (the second column of Table 2.1) in a column named POPN. (This named column may still be in your workspace.) We will store the results in the column SMPL. Then proceed as follows.

Menu commands:

1 Choose Calc ➢ Random Data ➢ Sample From Columns...
2 Type 2 in the Sample [ ] rows from column(s): text box
3 Press [Tab] and enter POPN in the large text box.
4 Select the Store samples in: text box and enter SMPL.
5 Select [OK]

Session commands:

- Name an empty column SMPL if necessary
- Type SAMPLE 2 'POPN' 'SMPL' at the MTB > prompt and press [Enter]

This much stores the two sample values in the column SMPL and we want to compare the mean of these two sample values with the population mean μ.

The mean of this population of salaries is

$$\mu = \frac{\Sigma x}{N} = \frac{70 + 40 + 37 + 55 + 50}{5} = 50.4,$$

or \$50,400.

The sample mean, $\overline{x}$, of the two salaries obtained can be determined using the MEAN command in the usual way. See Printout 7.1 for the result, which may differ from yours.

PRINTOUT 7.1
Sample mean

```
Mean of SMPL    =        47.500
```

The printout shows that the sample mean of the two randomly selected salaries is given by $\overline{x} = 47.5$ (that is, \$47,500). Comparing the sample mean of $\overline{x} = 47.5$ to the population mean of $\mu = 50.4$, we see that in this case the sampling error made in estimating μ by $\overline{x}$ is 2.9 (that is, \$2,900). ■

PROBLEM 7.1 Repeat Example 7.1 three times. [Note: It is possible that you may obtain the same random sample more than once.] ◇

7.2 The sampling distribution of the mean **E-7.3:I-7.3**

An important tool in statistical analysis is related to knowledge of the probability distribution of the mean, $\overline{X}$. The probability distribution of this random variable is so important that it is given a special name—the *sampling distribution of the mean.*

The sampling distribution of the mean depends on the distribution of the population being sampled and on the sample size. In general, it is not possible to obtain the sampling distribution of the mean *exactly.* Even though it is generally not possible to determine the sampling distribution of the mean exactly, there are mathematical relationships that permit us to obtain the sampling distribution of the mean approximately. It is helpful to distinguish between the case where the population being sampled is normally distributed and the case where it may not be so. Let us first consider normally distributed populations.

NORMALLY DISTRIBUTED POPULATIONS

Suppose the population under consideration is *normally distributed.* Then, although it is by no means obvious, the random variable $\overline{X}$ is also normally distributed. Specifically we have the following result:

KEY FACT 7.1 The sampling distribution of the mean for normal populations

Suppose a random sample of size n is to be taken from a *normally distributed* population with mean μ and standard deviation σ. Then the random variable $\overline{X}$ is also *normally distributed* and has mean $\mu_{\overline{x}} = \mu$ and standard deviation $\sigma_{\overline{x}} = \sigma/\sqrt{n}$. In other words, if the population being sampled is normally distributed, then probabilities for $\overline{X}$ are equal to areas under the normal curve with parameters μ and $\sigma/\sqrt{n}$.

An intuitive interpretation of this Key Fact is as follows: Suppose we compute the sample mean, $\overline{X}$, for each possible sample of size n from a normally distributed population. Then the histogram of all of those $\overline{x}$-values will be bell-shaped. Let us now illustrate this intuitive interpretation with the aid of Minitab.

EXAMPLE 7.2 Illustrates the sampling distribution of the mean for normal populations

The ages of farm operators in the United States are assumed to be *normally distributed* with a mean of $\mu = 50$ years and a standard deviation of $\sigma = 8$ years. Apply Minitab to illustrate the sampling distribution of the mean for this population of ages when $n = 3$.

SOLUTION Since the population is normally distributed, the random variable $\overline{X}$ is also normally distributed and has mean $\mu_{\overline{x}} = \mu = 50$, the same as the population mean. However, the standard deviation is $\sigma_{\overline{x}} = \sigma/\sqrt{n} = 8/\sqrt{3} = 4.62$. Thus the set of all possible $\overline{x}$-values obtained from all possible samples of size $n = 3$ has a normal distribution with mean $\mu_{\overline{x}} = 50$ and standard deviation $\sigma_{\overline{x}} = 4.62$.

Unfortunately, we cannot get all the $\overline{x}$-values since, at least theoretically, there are an infinite number of them. However, we can make plausible the fact that the set of all $\overline{x}$-values is normally distributed by obtaining a large number of samples of size three and showing that the histogram of their $\overline{x}$-values is roughly bell-shaped.

In Lesson 6 you learned how Minitab can be used to simulate random sampling from a normally distributed population. For our population of ages, we know that $\mu = 50$ and $\sigma = 8$. We will employ Minitab to (1) take 1000 samples of size $n = 3$, (2) compute $\overline{x}$ for each of the 1000 samples, and (3) draw a histogram of the 1000 $\overline{x}$-values. The resulting histogram should be roughly bell-shaped.

We now proceed to accomplish (1)–(3) as follows.

Menu commands:

1 Choose Calc ➤ Random Data ➤ Normal...

2 Type 1000 in the Generate [ ] rows of data text box

3 Select the Store in column(s): text box and type C101-C103 (any three successive empty columns will do)

4 Select the Mean: text box and type 50

5 Select the Standard deviation: text box and type 8

6 Select [OK]

Session commands:

- Type RANDOM 1000 C101-C103; at the MTB > prompt and press Enter↵
- Type NORMAL 50, 8. at the SUBC> prompt and press Enter↵

These commands tell Minitab to place a total of 3000 random observations from the population into columns C101–C103, with 1000 observations in each column. Then our first random sample of size $n = 3$ is the first row of columns C101–C103, our second random sample of size $n = 3$ is the second row of columns C101–C103, and so on.†

Second, we compute the sample mean, $\overline{x}$, for each of the 1000 samples. We use a command that computes row means instead of column means as follows.

Menu commands:

1 Choose Calc ➤ Row Statistics...
2 Select the Mean option button under Statistic
3 Select the Input variables: text box and type C101-C103
4 Select the Store result in: text box, and type XBARS
5 Select OK

Session commands:

- Name an empty colum XBARS
- Type RMEAN C101-C103 'XBARS' at the MTB > prompt and press Enter↵

These commands instruct Minitab to compute the means of the 1000 *rows* of data located in C101–C103 and to place those means in XBARS. Consequently, the 1000 $\overline{x}$-values are now in XBARS.

Finally, to obtain a histogram of the 1000 $\overline{x}$-values with a superimposed normal curve, proceed as follows.

Menu commands:

1 Choose Stat ➤ Basic Statistics ➤ Display Descriptive Statistics...
2 Enter XBARS in the Variables: text box
3 Select the Graphs... button
4 Select the Histogram of data, with normal curve check box
5 Select OK twice

Session commands:

- Type HISTOGRAM 'XBARS' at the MTB > prompt and press Enter↵ (There is no normal curve option.)

The results are displayed in Figure 7.1 on the following page.

† Note the use of the command RANDOM rather than SAMPLE.

FIGURE 7.1
Sampling distribution of the mean; normal population

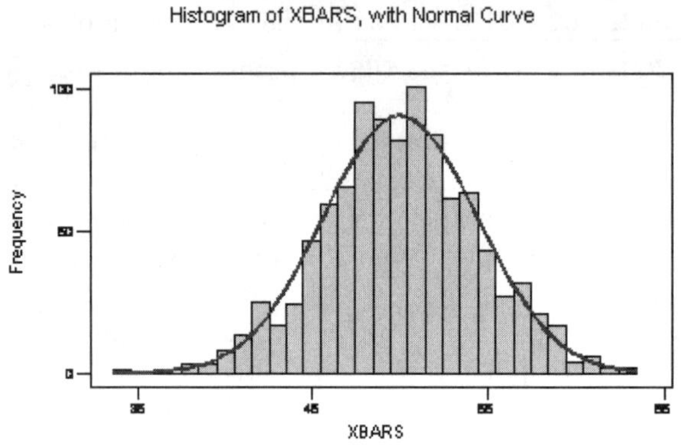

As we expected, the histogram of the 1000 $\overline{x}$-values is roughly bell-shaped. Your display may differ from this one because of the particular random samples you obtained. If we took, say, 5000 samples of size $n = 3$ and obtained a histogram of the 5000 $\overline{x}$-values, then that histogram would be even more bell-shaped. ■

PROBLEM 7.2 Refer to Example 7.2. Use Minitab to

a) take 500 random samples of size $n = 9$ from the population of ages.
b) compute $\overline{x}$ for each sample obtained in part (a).
c) draw a histogram of the 500 $\overline{x}$-values from part(b) with a superimposed normal curve. ◇

GENERAL POPULATIONS

We have seen that if the population being sampled is normally distributed, then so is the sampling distribution of the mean (that is, the probability distribution of the sample mean, $\overline{X}$). Remarkably, when the sample size is relatively large, the sampling distribution of the mean is approximately a normal distribution, even if the population being sampled is not normally distributed. This extraordinary fact is called the *central limit theorem*:

KEY FACT 7.2 The central limit theorem

For a relatively large sample size, the random variable $\overline{X}$ is approximately *normally distributed,* **regardless of how the population is distributed. The approximation becomes better and better with increasing sample size.**

In general, the more "nonnormal" the population, the larger the sample size must be in order to use normal curve areas to determine probabilities for $\overline{X}$. As a rule of thumb, we will consider sample sizes of 30 or more ($n \geq 30$) large enough. The following statement summarizes the result.

KEY FACT 7.3 The sampling distribution of the mean for general populations

Suppose a random sample of size $n \geq 30$ is to be taken from a population with mean μ and standard deviation σ. Then regardless of the distribution of the population, the random variable $\overline{X}$ is approximately *normally distributed* **and has mean $\mu_{\overline{x}} = \mu$ and standard deviation $\sigma_{\overline{x}} = \sigma/\sqrt{n}$. That is, probabilities for $\overline{X}$ are approximately equal to areas under the normal curve with parameters μ and $\sigma/\sqrt{n}$.**

In the next example we will see how Minitab can be used to illustrate the central limit theorem and the sampling distribution of the mean for a population that is *not* normally distributed.

EXAMPLE 7.3 Illustrates the central limit theorem and sampling distribution of the mean for nonnormal populations

In 1980 a telephone company in the Southwest undertook a study of telephone usage in Phoenix, Arizona. The results indicate that the durations of telephone conversations by residential customers in Phoenix have an *exponential distribution* with a parameter μ roughly equal to 4 minutes. This means the following: Let the population under consideration be the durations, in minutes, of all possible telephone conversations by Phoenix residential customers. Then percentages for that population are equal to areas under the curve

$$y = \frac{1}{4}e^{-x/4}.$$

This curve is pictured in Figure 7.2.

FIGURE 7.2
Exponential curve for duration of conversation

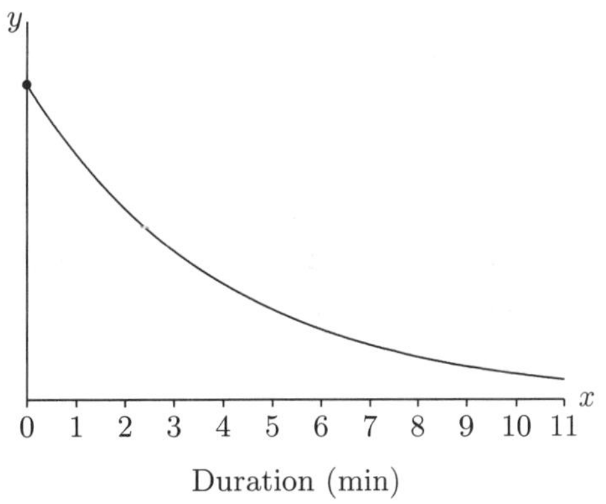

Clearly this population is not normally distributed. Nonetheless, according to the central limit theorem, the random variable $\overline{X}$ is approximately normally distributed when the sample size is relatively large. Use Minitab to illustrate this fact.

SOLUTION We first explain how Minitab can be used to simulate a random sample from an exponential distribution. The procedure is analogous to that for a normal distribution. In fact, we still employ the command RANDOM, but for an exponential distribution we use a subcommand called EXPONENTIAL with a specification of a single parameter, the mean, μ, of the distribution.

The central limit theorem says that the sampling distribution of the mean (that is, the probability distribution of $\overline{X}$) is approximately normally distributed when the sample size is $n = 30$, as it is in this example. Here we have a population that is badly skewed to the right. Nevertheless, the theorem assures us that the distribution of averages based on a sample size of $n = 30$ will not be as skewed as the population. Indeed, the distribution will resemble that of a normal distribution.

Let us now employ Minitab to (1) take, say, 250 samples of size $n = 30$, (2) compute $\overline{x}$ for each of the 250 samples obtained, and (3) draw a histogram of the 250 $\overline{x}$-values. We proceed as we did in Example 7.2.

Menu commands:

1 Choose Calc ➤ Random Data ➤ Exponential...

2 Type 250 in the Generate [ ] rows of data text box

3 Select the Store in column(s): text box and type C111-C140 (any thirty successive empty columns will do)

4 Select the Mean: text box and type 4

5 Select [OK]

Session commands:

- Type RANDOM 1000 C111-C140; at the MTB > prompt and press [Enter ↵]
- Type EXPONENTIAL 4. at the SUBC> prompt and press [Enter ↵]

Next, compute the sample mean, $\overline{x}$, for each of the 250 samples as before.

Menu commands:

1 Choose Calc ➤ Row Statistics...

2 Select the Mean option button under Statistic

3 Select the Input variables: text box and type C111-C140

4 Select the Store result in: text box, and type AVGTIME

5 Select [OK]

Session commands:

- Name an empty colum AVGTIME
- Type RMEAN C111-C140 'AVGTIME' at the MTB > prompt and press [Enter ↵]

Finally, we obtain a histogram of the 250 $\overline{x}$-values with a superimposed normal curve.

Menu commands:

1 Choose Stat ➤ Basic Statistics ➤ Display Descriptive Statistics...

2 Enter **AVGTIME** in the Variables: text box

3 Select the Graphs... button

4 Select the Histogram of data, with normal curve option

5 Select OK twice

Session commands:

- Type HISTOGRAM 'AVGTIME' at the MTB > prompt and press Enter (There is no normal curve option.)

The results (yours will differ slightly) are displayed in Figure 7.3.

FIGURE 7.3 Sampling distribution of the mean; exponential population

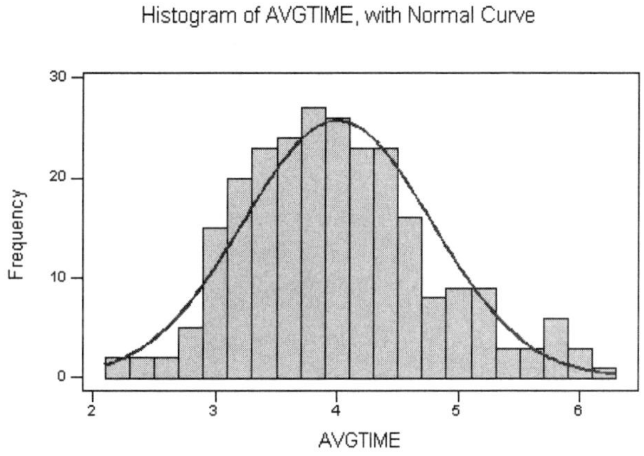

The histogram in Figure 7.3 seems to confirm that the sampling distribution is approximately normally distributed even though the population is not normally distributed. Selecting a larger sample size would make the resemblance even more dramatic. ■

PROBLEM 7.3

Refer to Example 7.3.

a) Use Minitab to (1) take 500 samples of size $n = 2$ from the population of phone conversation durations, (2) compute $\overline{x}$ for each of the 500 samples obtained, and (3) draw a histogram of the 500 $\overline{x}$-values. Interpret your results in light of the central limit theorem.

b) Repeat part (a) for samples of size $n = 10$. ◇

LESSON 8

Confidence Intervals for One Population Mean

GENERAL OBJECTIVE

We now begin our study of how Minitab can be used in inferential statistics. In this lesson, we will employ Minitab to obtain *confidence intervals* for a population mean, μ. Specifically, we will learn the Minitab commands that permit us to perform Procedures 8.1 and 8.2 in *IS/6e* and *ES/5e* with the computer.

LESSON OUTLINE

8.1 Confidence intervals for one population mean; σ known

E-8.2:I-8.2

Procedure 8.1 in *IS/6e* and *ES/5e* gives a step-by-step method for obtaining a confidence interval for a population mean, μ, when σ is known. Minitab has a command called ZINTERVAL that will perform Procedure 8.1 automatically for us. Example 8.4 of *IS/6e* and *ES/5e* illustrates the application of Procedure 8.1 to determine a confidence interval for the mean age of all people in the civilian labor force. Let us use that example along with the ZINTERVAL command to obtain that confidence interval.

EXAMPLE 8.1 *Illustrates computing a z-confidence interval*

The U.S. Bureau of Labor Statistics collects information on the ages of people in the civilian labor force and publishes data on the ages of people in the civilian labor force. Suppose $n = 50$ such people are randomly selected and that their ages are as displayed in Table 8.1 (WeissStats CD:Tb08-03). Assume $\sigma = 12.1$ years.

TABLE 8.1
Ages of 50 randomly selected people in the civilian labor force

22	58	40	42	43	32	34	45	38	19
33	16	49	29	30	43	37	19	21	62
60	41	28	35	37	51	37	65	57	26
27	31	33	24	34	28	39	43	26	38
42	40	31	34	38	35	29	33	32	33

Use Minitab to find a 95% confidence interval for the mean age, μ, of all people in the civilian labor force.

SOLUTION The command ZINTERVAL requires the user to specify the population standard deviation, σ. So if the latter is unknown, you will not be able to use this procedure. To begin, enter the sample data of Table 8.1 into a column named AGE in the usual manner. Then proceed as follows.

Menu commands:

1 Choose Stat ➤ Basic Statistics ➤ 1-Sample Z...

2 Enter AGE in the Variables: text box

3 Select the Confidence interval option button

4 Select the Level: text box and type 95

5 Select the Sigma: text box and type 12.1

6 Select [OK]

Session commands:

- Type ZINTERVAL 95 12.1 'AGE' at the MTB > prompt and press [Enter]

Printout 8.1 on the following page shows the results.

PRINTOUT 8.1
Minitab
confidence interval

```
The assumed sigma = 12.1

Variable    N      Mean    StDev  SE Mean          95.0 % C.I.
AGE        50     36.38    11.07     1.57  (   33.03,    39.73)
```

The first line of the output displays the value used for the population standard deviation, σ: `The assumed sigma = 12.1`. Then we find the sample size, sample mean, sample standard deviation, and standard error of the mean ($\sigma/\sqrt{n}$), basic items you will find when you execute the `DESCRIBE` command. The final item is the confidence interval.

Hence a 95% confidence interval for μ is from 33.03 to 39.73. We can be 95% confident that the mean age, μ, of all people in the civilian labor force is somewhere between 33.03 and 39.73 years. ■

PROBLEM 8.1

The U.S. National Center for Health Statistics estimates mean weights and heights of U.S. adults by age and sex. Suppose that 40 women, 5 feet 4 inches tall and aged 18–24, are randomly selected and that their weights, in pounds, are as shown here.

140	136	147	138	143	122	115	125
136	152	130	134	150	153	148	132
116	159	128	136	134	126	120	146
131	167	145	132	138	137	115	145
154	139	139	147	123	154	127	116

Use Minitab to find a 90% confidence interval for the mean weight, μ, of all U.S. women 5 feet 4 inches tall and in the age group 18–24. Use `WEIGHT` for the name of the storage column. Assume $\sigma = 12.8$. ◇

8.2 Confidence intervals for one population mean; σ unknown

E-8.4:I-8.4

We can also use Minitab to obtain a confidence interval for a population mean, μ, when the population being sampled is normally distributed and the standard deviation σ is unknown. The steps are summarized as Procedure 8.2 in *IS/6e* and *ES/5e*. The appropriate Minitab command is called TINTERVAL (the first "T" in the name stands for the t-distribution).

The formulas involve finding t-values corresponding to a specified area under a t-curve, that is, percentile points and we investigate that problem first. Like the standard normal distribution, the t-distribution is symmetric and centered about the value of 0.

FINDING THE t-VALUE FOR A SPECIFIED AREA

Minitab has a subcommand to the INVCDF command called t, with a specification of degrees of freedom (df). That means you can find the t-value corresponding to any area to the left of that value for any t-curve.

EXAMPLE 8.2 Illustrates finding the t-value for a specified area using INVCDF

For a t-curve with 13 degrees of freedom, find the t-value with area 0.05 to its right using Minitab.

SOLUTION Since the area to the left is 0.95, we proceed as follows.

Menu commands:

1. Choose Calc ➤ Probability Distributions ➤ t...
2. Select the Inverse cumulative probability option button
3. Select the Degrees of freedom: text box and type 13
4. Select the Input constant: radio button, press [Tab], and type 0.95
5. Select [OK]

Session commands:

- Type `INVCDF 0.95` at the `MTB >` prompt and press [Enter]
- Type `t 13.` at the `SUBC>` prompt and press [Enter]

Printout 8.2 shows the result.

PRINTOUT 8.2 Output of INVCDF with the t subcommand

```
Student's t distribution with 13 DF

 P( X <= x)          x
    0.9500      1.7709
```

Thus 1.7709 is the required t-value; that is, $t_{0.05} = 1.7709$. ■

PROBLEM 8.2

For a t-curve with df=8, use Minitab to find the following t-values.

a) The t-value with area 0.10 to its right.

b) $t_{0.005}$.

c) The t-value with area 0.01 to its left. ◇

CONFIDENCE INTERVALS: t-INTERVAL

A step-by-step method of finding a confidence interval for the mean when σ is unknown is outlined in Procedure 8.2 in *IS/6e* and *ES/5e*. The following example shows how to determine that confidence interval by employing the TINTERVAL command in Minitab.

EXAMPLE 8.3 *Illustrates Procedure 8.2 for computing a t-interval*

The U.S. Federal Bureau of Investigation (FBI) compiles data on robbery and property crimes and publishes the information in *Population-at-Risk Rates and Selected Crime Indicators*. A sample of last year's pick-pocket offenses yields the values lost shown in Table 8.2 (WeissStats CD:Tb08-05).

TABLE 8.2 Value lost ($) for a sample of 25 pick-pocket offenses

447	207	627	430	883
313	844	253	397	214
217	768	1064	26	587
833	277	805	653	549
649	554	570	223	443

Use Minitab to obtain a 95% confidence interval for the mean value lost, μ, of all last year's pick-pocket offenses.

SOLUTION We need to be reasonably sure that the population of values lost is normally distributed if we are to apply the methods of this section, for the sample size is not large. First, store the sample data of Table 8.2 into a column named `VALUE` in the usual way. Next, we convert the data to Nscores (as in Lesson 6, page 104) and examine the resulting normal probability plot using the methods of Section 6.4. The result is displayed in Figure 8.1.

FIGURE 8.1 Normal probability plot of value-lost data in Table 8.2

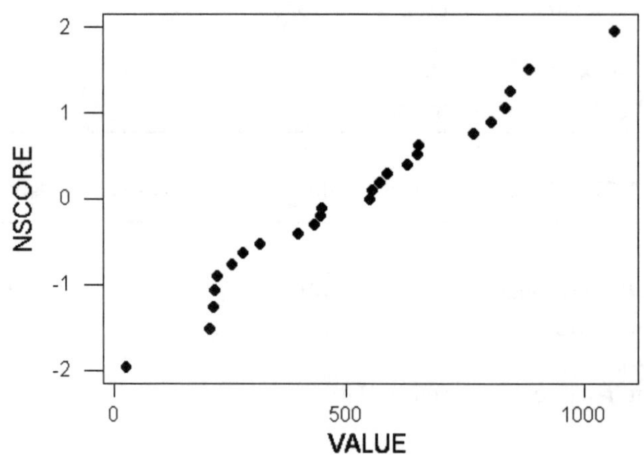

The normal probability plot in Figure 8.1 shows no outliers and falls roughly in a straight line. This suggests that it is reasonable to assume that the population

of values is normally distributed. That being the case, we employ TINTERVAL to find the 95% confidence interval. Proceed as follows.

Menu commands:

1 Choose Stat ➤ Basic Statistics ➤ 1-Sample t...

2 Enter the variable **VALUE** in the Variables: text box

3 Select the Level: text box and type 95

4 Select [OK]

Session commands:

- Type `TINTERVAL 95 'VALUE'` at the `MTB >` prompt and press [Enter↵]

(If you omit the percentage specification, Minitab will compute a 95% confidence interval automatically.)

Printout 8.3 shows the output that results.

PRINTOUT 8.3
Minitab output for TINTERVAL

```
Variable      N      Mean     StDev  SE Mean          95.0 % CI
VALUE        25     513.3     262.2     52.4  (   405.1,    621.6)
```

The output first displays the sample size, sample mean, sample standard deviation, and estimated standard error of the mean $(s/\sqrt{n})$ as in the DESCRIBE command. Then comes the confidence interval. From this last item we see that a 95% confidence interval for μ is from 405.1 to 621.6. We can be 95% confident that the mean value lost, is somewhere between $405.10 and $621.60. ■

PROBLEM 8.3 According to the R.R. Bowker Company of New York, the mean annual subscription rate to law periodicals was $97.33 in 1995. In 1996, a random sample of $n = 12$ law periodicals yields current annual subscription rates, to the nearest dollar, as displayed in the following table.

106	122	120	123
118	114	138	131
128	124	119	130

Test the 1996 annual subscription rates to law periodicals for normality. If you are satisfied that rates are normally distributed, use Minitab to obtain a 95% confidence interval for this year's mean annual subscription rate, μ, for law periodicals. Use **SUBRATES** as the name of the storage column. ◇

LESSON 9

Hypothesis Tests for One Population Mean

GENERAL OBJECTIVE

In Lesson 8 we discovered how Minitab can be used to obtain confidence intervals for one population mean, μ. Now we will examine the Minitab commands that perform *hypothesis tests* for μ. These commands correspond to various procedures in *IS/6e* and *ES/5e*.

LESSON OUTLINE

9.1 Some preliminaries

E-9.2:I-9.2

Before commencing our study of how Minitab can be used to perform hypothesis tests for one population mean, μ, we need to consider some preliminaries. We begin by discussing how the alternative hypothesis is specified when using Minitab.

SPECIFYING THE ALTERNATIVE HYPOTHESIS

Recall that typically there are two hypotheses in a hypothesis test—the *null hypothesis* ($\boldsymbol{H}_0$) and the *alternative hypothesis* ($\boldsymbol{H}_a$). The goal in a hypothesis test is to decide whether or not the null hypothesis should be rejected in favor of the alternative hypothesis based on the observed data.

In hypothesis tests concerning a single population mean, μ, the null hypothesis generally specifies a *single value* for that parameter. In other words, the null hypothesis is of the form

$$H_0 : \mu = \mu_0,$$

where μ_0 is some specified number.

On the other hand, there are *three* possibilities for the alternative hypothesis. It can be of the form $H_a : \mu \neq \mu_0$, $H_a : \mu < \mu_0$, or $H_a : \mu > \mu_0$. The hypothesis test is called *two-tailed* in the first case, *left-tailed* in the second case, and *right-tailed* in the third case. Table 9.1 summarizes this discussion.

TABLE 9.1 Possible alternative hypotheses for hypothesis tests concerning a single population mean

Type	Form
Two-tailed	$H_a : \mu \neq \mu_0$
Left-tailed	$H_a : \mu < \mu_0$
Right-tailed	$H_a : \mu > \mu_0$

When we use Minitab to perform a hypothesis test for a population mean, μ, we must indicate which type of test is to be performed—two-tailed, left-tailed, or right-tailed. Minitab is preset for a two-tailed test when we use session commands. If we want to perform a left-tailed test, we type `ALTERNATIVE -1` as a subcommand. If we want to perform a right-tailed test, we type `ALTERNATIVE 1`. Similar conventions exist as choices in the menu commands. See Table 9.2.

TABLE 9.2 Subcommands for specifying the alternative hypothesis in session commands

Type of test	Subcommand
Two-tailed	ALTERNATIVE 0
Left-tailed	ALTERNATIVE − 1
Right-tailed	ALTERNATIVE 1

A *classical* test of a hypothesis is conducted by finding an appropriate *test statistic* and a corresponding *rejection region*, given a level of significance. In our applications, this region is determined by finding percentile points under an appropriate probability distribution. These values are called *critical values* and a proper comparison of the value of the test statistic to these critical values leads to the decision to reject or not reject H_0. The hypothesis is rejected (at that level of significance) only when the value of the test statistic falls in this rejection region. This is referred to as the *critical value approach* in your textbook. See Procedure 9.1 and the related discussion in *IS/6e* and *ES/5e*. With modern technology, that process can be shortened somewhat with the notion of a *P*-value.

P-VALUES

The *P-value* for a hypothesis test is the smallest significance level at which the null hypothesis can be rejected with the observed sample data. When we use Minitab to perform a hypothesis test, the *P*-value is usually printed out. To test a hypothesis using the *P*-value, we have the following decision criterion.

KEY FACT 9.1 Decision criterion for a hypothesis test using the *P*-value

If the *P*-value is less than or equal to the specified significance level, α, then reject the null hypothesis. Otherwise, do not reject the null hypothesis. Your textbook calls this the *P-value approach* to hypothesis testing.

(Note: While *P*-value has a wider application than merely a way of conducting a classical test of hypothesis, its use in this manual is restricted to its compliance with Key Fact 9.1.*)*

RELATIONSHIP BETWEEN HYPOTHESIS TESTS AND CONFIDENCE INTERVALS

The equivalence of testing a hypothesis and observing whether or not the appropriate confidence interval contains the hypothesized value is discussed briefly just before the Exercises for Section 9.3 of *IS/6e* and *ES/5e*. The details are left for the exercises. The discussion there is restricted to a two-tailed test, however. With some modification, the same equivalence may be found for one-tailed tests as well.

Let us examine the Right-tailed test discussed in Procedure 9.1 of *IS/6e* and *ES/5e*. You see there that you *FAIL* to reject the null hypothesis when

$$\frac{\overline{x} - \mu_0}{\sigma/\sqrt{n}} < z_\alpha$$

and, otherwise, you reject the null hypothesis. But it is easy to verify that this inequality is equivalent to stating

$$\overline{x} - z_\alpha \cdot \sigma/\sqrt{n} < \mu_0$$

However, if you will examine the formulas in Procedure 8.1 in *IS/6e* and *ES/5e*, you will see that the confidence limits for a confidence interval for μ with a confidence level of $1 - 2\alpha$ would be from

$$\overline{x} - z_\alpha \cdot \sigma/\sqrt{n} \quad \text{to} \quad \overline{x} + z_\alpha \cdot \sigma/\sqrt{n}$$

We call the interval running from

$$\overline{x} - z_\alpha \cdot \sigma/\sqrt{n} \quad \text{to} \quad \infty$$

a *lower one-sided confidence interval for* μ with confidence level $1 - \alpha$. Similarly, we call the interval running from

$$-\infty \quad \text{to} \quad \overline{x} + z_\alpha \cdot \sigma/\sqrt{n}$$

an *upper one-sided confidence interval for* μ with confidence level $1 - \alpha$.

To summarize, for the Right-tailed test you would *FAIL* to reject the null hypothesis when the value of μ_0 falls inside the *LOWER* one-sided confidence interval; otherwise reject the null hypothesis (namely, when μ_0 is outside the interval). Similarly, for the Left-tailed test you would *FAIL* to reject the null hypothesis when the value of μ_0 falls inside the *UPPER* one-sided confidence interval; otherwise reject the null hypothesis (namely, when μ_0 is outside the interval).

Notice the match Right-tail test to Lower one-sided interval and Left-tail test to Upper one-sided interval. It happens that this equivalence is valid in practically all of the procedures in this manual when both hypothesis tests and confidence intervals are available. We will bring these matters to your attention in the examples and list it here as another key fact outside of your textbook.

KEY FACT 9.2 Decision criterion for a hypothesis test using confidence intervals

Once the proper match between alternatives and confidence intervals is made, then, for a specified significance level, α, reject the null hypothesis when the specified value lies outside the confidence interval. Otherwise, do not reject the null hypothesis.

Hypothesis testing using the three approaches we have outlined are equivalent in the sense that they all lead to exactly the same action.† That is, if one procedure leads to the final action of rejecting (or failing to reject) the hypothesis, the others will result in the same action. Yet the approaches have a slightly different perspective in terms of application.

Thus, in hypothesis testing, *for a given level of significance* α, the P-value tells at a glance, all significance levels at which the hypothesis would be rejected (those smaller than α) and those levels at which you would not reject the hypothesis (those greater than α).

On the other hand, *for a given level of significance* α, the confidence interval reveals at a glance all those choices for H_0 for which the hypothesis would be rejected (those outside the confidence interval) and those choices for which you would not reject the hypothesis (those inside the confidence interval).

Another view of this difference is that if you change the level of significance, α, the P-value does not change but the confidence interval does. If you change the hypothesis, H_0, you wish to test, the confidence interval does not change but the P-value does.

† For an exception, see Problem 12.3.

The critical value approach on the other hand is fixed both to the level of significance given and the null hypothesis. A change in either one of these factors requires computing new critical values. That is, *for a given level of significance* α and a given null hypothesis, H_0, the procedure tells you whether or not to reject *that* hypothesis at *that* level of significance. It reveals little about other hypotheses or levels of significance.

Starting with R13, Minitab computes one-sided intervals and displays the appropriate bounds in nearly all of their testing procedures. For earlier releases, you can find the appropriate limit for significance level α by computing with confidence level $1 - 2\alpha$ and ignoring one of the limits in the output. For example, if you are performing a left-tailed or right-tailed test at the 5% level of significance, compute a 90% confidence interval to find 95% one-sided confidence bounds. We will bring this to your attention in the examples.

9.2 One-sample z-test for a population mean **E-9.4:I-9.5**

We can use Minitab to perform the one-sample z-test (Procedures 9.1 and 9.2 in *IS/6e* and *ES/5e*), a hypothesis test for a population mean, μ, when the standard deviation, σ, is known. At the same time, we can illustrate the P-value approach and the equivalence to confidence intervals. The appropriate command for getting results in Minitab is called ZTEST. To explain the details, we will use the setting of an example in *IS/6e* and *ES/5e*.

EXAMPLE 9.1 *Illustrates the one-sample z-test*

Calcium is the most abundant and one of the most important minerals in the body. It works with phosphorus to build and maintain bones and teeth. According to the Food and Nutrition Board of the National Academy of Sciences, the recommended daily allowance (RDA) of calcium for adults is 800 mg (milligrams).

A nutritionist claims that the average person with an income below the poverty level gets less than the RDA of 800 mg. To test her claim, she obtains the daily intakes of calcium for a random sample of 35 people with incomes below the poverty level. The results are displayed in Table 9.3 (WeissStats CD:Tb09-06). Data are in milligrams.

TABLE 9.3
Daily calcium intakes

686	433	743	647	734	641
993	620	574	634	850	858
992	775	1113	672	879	609

At the 5% significance level, do the data provide sufficient evidence to conclude that the mean calcium intake of all people with incomes below the poverty level is less than the RDA of 800 mg per day? Use Minitab to perform the hypothesis test. Assume $\sigma = 188$ mg.

SOLUTION Let μ denote the mean daily intake of calcium of all people with incomes below the poverty level. The problem is to perform the hypothesis test

H_0: $\mu = 800$ mg (mean calcium intake is not less than the RDA)
H_a: $\mu < 800$ mg (mean calcium intake is less than the RDA)

at the 5% significance level ($\alpha = 0.05$). Note that the hypothesis test is left-tailed since there is a less-than sign ($<$) in the alternative hypothesis. This means we will need an upper one-sided 95% confidence interval.

Enter the sample data from Table 9.3 into a column named `CALCI` in the usual manner. Proceed as follows.

Menu commands:

1 Choose Stat ➢ Basic Statistics ➢ 1-Sample Z...
2 Enter `CALCI` in the Variables: text box
3 Select the Sigma: text box and type 188
4 Select the Test mean: text box and type 800
5 Click the Options button and select less than in the Alternative: text box
6 Select [OK] twice

Session commands:

- Type `ZTEST 800 188 'CALCI';` at the `MTB` > prompt and press [Enter↵]
- Type `ALTERNATIVE -1.` at the `SUBC>` prompt and press [Enter↵]

The resulting output is displayed in Printout 9.1. The output first displays a statement of the null and alternative hypotheses for the hypothesis test: `Test of mu = 800.0 vs mu < 800.0`.

PRINTOUT 9.1
Minitab output
for ZTEST

```
Test of mu = 800 vs mu < 800
The assumed sigma = 188

Variable        N      Mean     StDev   SE Mean
CALCI          18     747.4     172.0      44.3

Variable    95.0% Upper Bound         Z       P
CALCI                   820.3     -1.19   0.118
```

Next, you are reminded of the value used for the population standard deviation, σ in the line `The assumed sigma = 188`. Following that, the output shows the sample size, sample mean, sample standard deviation, and standard error of the mean, much as in the output of the DESCRIBE command.

Following that summary, you see that the 95% upper one-sided interval for μ is $(-\infty, 820.3)$. *(Note:* For an earlier release, you may compute a 90% confidence interval by typing `ZINT 90 188 'CALCI'` at the `MTB` > prompt and discover that this interval is given by $674.5 < \mu < 820.3$. Use only the upper bound.) Using the `INVCDF 0.05` command in a session window, you can verify that the single critical value is $-z_{0.05} = -1.6449$ giving a rejection region of $(-\infty, -1.6449)$. After this follows `Z`, the value of the test statistic

$$z = \frac{\overline{x} - \mu_0}{s/\sqrt{n}} = \frac{747.314 - 800}{262.23/\sqrt{35}} = \frac{-52.686}{44.325}$$

which is equal to $z = -1.19$. The final entry shown in the output in Printout 9.1 is the `P-Value`.

Now we are ready to test the null hypothesis by the three methods we have presented. The fastest and easiest method is to observe that the computed P-value of 0.118 is greater than the level of significance $\alpha = .05$. Secondly, the hypothesized value of $\mu_0 = 800$ is less than the upper bound of $u = 820.3$, that is, lies in the confidence interval. Finally, since $-1.19 > -1.6449$, the value of the test statistic is not in the rejection region. In any case, you cannot reject the null hypothesis. The data do not provide sufficient evidence to conclude that the mean calcium intake of all people with incomes below the poverty level is less than the RDA of 800 mg per day.

However, you actually know more than this. You also know that no hypothesized value less than 820.3 would have been rejected by these data—at the 5% level of significance. Moreover, for this hypothesized value of 800, the level of significance would have to be larger than 11.8% before these data would have rejected this hypothesis. Try it. Test the hypothesis $\mu_0 = 821$ (or higher) at the 5% level of significance and see that you would reject also. Notice with the choice 821, the P-value changed to 0.048, a value less that 5%. Also the value of the test statistic is now -1.66 and that falls in the rejection region. ■

PROBLEM 9.1 *The World Almanac, 1985* reports that the average travel time to work in 1980 for residents of South Dakota was 13 minutes. For this year, a random sample of 35 travel times for South Dakota residents yielded the data below (in minutes).

29	40	0	12	10	6	41
25	21	5	4	19	2	7
10	8	3	6	52	4	12
0	33	6	2	17	21	8
38	2	13	8	14	11	2

Use Minitab to test, at the 5% significance level, whether this year's mean travel time to work for South Dakota residents appears to have *changed* from the 1980 mean of 13 minutes. Assume $\sigma = 11.6$ minutes and use `TRAVTIME` for the name of the column where you store the data. ◇

9.3 One-sample t-test for a population mean **E-9.5:I-9.6**

Procedure 9.3 in *IS/6e* and *ES/5e*is a step-by-step method for performing a hypothesis test concerning a population mean, μ, when the population being sampled is normally distributed or the sample size is large. We can use Minitab to carry out such a hypothesis test. The appropriate command is called TTEST. The next example explains in detail how to apply this procedure.

EXAMPLE 9.2 *Illustrates the t-test for a mean*

The U.S. Energy Information Administration surveys households to obtain data on residential energy consumption and expenditures. Results of the surveys can be found in *Residential Energy Consumption Survey: Consumption and Expenditures*. According to that publication, the mean residential energy expenditure of all American families was \$1282 in 1993. That same year, 36 randomly selected upper-income families reported the energy expenditures shown in Table 9.4.

TABLE 9.4 Energy expenditures (\$) for 36 households living in single-family detached homes

2016	1509	1658	1359	1564	1808	1155	1948	1162
1529	956	1284	2224	1549	1751	1408	1124	1083
1406	1370	1719	1647	1151	1735	1421	1403	1134
1877	1296	1734	1275	1341	932	1564	1012	1309

At the 5% significance level, do the data indicate that, in 1993, upper-income families spent more, on the average, for energy than the national average of \$1282? Use Minitab first to perform the hypothesis test.

SOLUTION First, store the sample data of Table 9.4 into a column named ENERGY in the usual way. Let μ denote the mean energy expenditure of all upper-income families in 1993. The problem is to perform the hypothesis test

$$H_0\colon \mu = \$1282 \text{ (mean was not greater than the national mean)}$$
$$H_a\colon \mu > \$1282 \text{ (mean was greater than the national mean)}$$

at the 5% significance level ($\alpha = 0.05$). Note that the hypothesis test is right-tailed since there is a greater-than sign (>) in the alternative hypothesis.

Since the standard deviation is unknown, we need to apply the t-test. Also, since the sample size is large, we can apply Minitab's procedure for carrying out the t-test without testing the data for normality. Proceed as follows.

Menu commands:

1 Choose Stat ➤ Basic Statistics ➤ 1-Sample t...
2 Enter ENERGY in the Variables: text box
3 Select the Test mean: text box and type 1282
4 Click the Options button and select greater than in the Alternative: text box

5 Select OK twice

Session commands:

- Type TTEST 1282 'ENERGY'; at the MTB > prompt and press Enter
- Type ALTE 1. at the SUBC> prompt and press Enter

Printout 9.2 displays the results.

PRINTOUT 9.2
Minitab output for a *t*-test

```
Test of mu = 1282 vs mu > 1282

Variable            N      Mean     StDev   SE Mean
ENERGY             36    1455.9     310.2      51.7

Variable       95.0% Lower Bound         T       P
ENERGY                    1368.6      3.36   0.001
```

On the first line of the output, we find a statement of the null and alternative hypotheses: Test of mu = 1282.0 vs mu > 1282.0. Then we find the sample size, sample mean, sample standard deviation, and estimated standard error of the mean. Following that is the 95% lower confidence bound yielding a lower one-sided confidence interval $(1368.6, \infty)$. *(Note:* If you have an earlier release, you may compute a 90% confidence interval by typing TINT 90 'ENERGY' at the MTB > prompt and discover that this interval is given by $1386.6 < \mu < 1543.3$. Use only the lower bound.*)* Next comes the value of the test statistic, 3.36. Using the session command INVCDF .95; with a t 35. subcommand, the single critical value is 1.6896 giving a rejection region of $(1.690, \infty)$. The final entry in Printout 9.2 shows the *P*-value of the hypothesis test, 0.001.

To test the null hypothesis, first observe that the computed *P*-value is less than the level of significance $\alpha = .05$. Since $3.36 > 1.6896$, the value of the test statistic lies inside the rejection region. The hypothesized value of 1282 is NOT included in the confidence interval. By all three methods, the null hypothesis should be rejected.

In other words, the data provide sufficient evidence to conclude that upper-income families spent more, on the average, for energy in 1987 than the national average of $1282. ■

PROBLEM 9.2 As reported by the College Entrance Examination Board in *National College-Bound Senior*, the mean verbal score on the Scholastic Assessment Test (SAT) in 1995 was 428 points out of a possible 800. A random sample of 25 verbal scores for last year yielded the following data.

344	494	350	376	313
489	358	383	498	556
379	301	432	560	494
418	483	444	477	420
492	287	434	514	613

At the 10% significance level, does it appear that last year's mean for verbal SAT scores is greater than the 1995 mean of 428 points? Store the data in a column named `SAT`. ◇

9.4 The Wilcoxon signed-rank test

I-9.7

The Wilcoxon signed-rank test can be used to perform a hypothesis test for a population median, η, as well as a population mean, μ. This is because the test assumes the population under consideration has a symmetric distribution. The test does not require that the distribution be normal or have any other specific shape. However, it does not apply to skewed populations and so some safeguards have to be taken to be sure that this is not the case.

A step-by-step method to perform a Wilcoxon signed-rank test for a population median, η, or a population mean μ, is given as Procedure 9.4 found in *IS/6e* . Minitab has a program that will carry out Procedure 9.4 for us. The appropriate command is called WTEST. There is a companion command called WINTERVAL that will compute a confidence interval for the population median. The next example shows how these commands are applied.

EXAMPLE 9.3 Illustrates the WTEST and WINTERVAL commands

A random sample of 10 Kansas families of four with intermediate budgets yielded the data on 1998 weekly food costs (in$) shown in Table 9.5 (WeissStats CD:Tb09-14).

TABLE 9.5
Weekly food costs

143	169	149	135	161
138	152	150	141	159

Do the data provide sufficient evidence to conclude that, in 1998, the mean weekly food cost for Kansas families of four with intermediate budgets was less than the national mean of $157? Use Minitab to perform the hypothesis test at the 5% level of significance. Also compute a 90% confidence interval for the national mean.

SOLUTION Let μ denote the 1998 mean weekly food cost for all Kansas families of four with intermediate budgets. Then the problem is to use the Wilcoxon signed-rank test to perform the hypothesis test

$$H_0\text{: } \mu = \$157 \text{ (mean weekly food cost is not less than \$157)}$$
$$H_a\text{: } \mu < \$157 \text{ (mean weekly food cost is less than \$157)}$$

at the 5% significance level. Note that the hypothesis test is left-tailed since there is a less-than sign (<) in the alternative hypothesis.

First, we need to be reasonably satisfied that the population is symmetric. In spite of the very small sample size, we will examine symmetry through a stem-and-leaf diagram. We begin by entering the sample data from Table 9.5 into a column named COST. Using the methods of Lesson 2, Printout 9.3, shows the results of the STEM command applied to the data.

PRINTOUT 9.3
Minitab stem-and-leaf plot for the weekly food cost data

```
Stem-and-leaf of COST       N  = 10
Leaf Unit = 1.0

    2   13 58
    4   14 13
    5   14 9
    5   15 02
    3   15 9
    2   16 1
    1   16 9
```

The stem-and-leaf diagram in the printout does not display any excessive violation of symmetry, and so we feel comfortable about applying the test. Proceed as follows to apply the Minitab procedure WTEST.

Menu commands:

1 Choose Stat ➤ Nonparametrics ➤ 1-Sample Wilcoxon...
2 Enter COST in the Variables: text box
3 Select the Test median option button
4 Press Tab, and type 157
5 Click the ▼ button in the Alternative: text box and select less than from drop-down list
6 Select OK

Session commands:

- Type WTEST 157 'COST'; at the MTB > prompt and press Enter
- Type ALTERNATIVE= -1. at the SUBC> prompt and press Enter

Printout 9.4 shows the resulting output.

PRINTOUT 9.4
Minitab output for WTEST

```
Test of median = 157.0 versus median  <  157.0

                   N for   Wilcoxon              Estimated
                N   Test  Statistic        P        Median
COST           10     10        9.0    0.033         149.5
```

The first line of the output displays the null and alternative hypotheses for the hypothesis test: Test of median = 157.0 versus median < 157.0. Below that we find several entries. The first entry, headed N, gives the sample size, which in this case is 10.

The next entry, headed `N for Test`, shows the number of data values in the sample that are not equal to the null hypothesis median of 157. Recall that for a Wilcoxon signed-rank test, if a data value is equal to the null hypothesis median, then that data value should be removed and the sample size reduced by 1. Since here the sample size is 10 and none of the data values in the sample are equal to 157, the sample size for the test (`N for test`) is also 10.

The third entry, headed `Wilcoxon Statistic`, provides the value of the test statistic

$$W = \text{sum of the positive ranks.}$$

So, we see that $W = 9$.

The last two entries, headed `P` and `Estimated Median`, respectively, give the P-value for the hypothesis test and a point estimate for the population median. Since the P-value of 0.033 is less than the designated significance level of $\alpha = 0.05$, we reject H_0. In other words, the data provide sufficient evidence to conclude that, in 1998, the mean weekly food cost for Kansas families of four with intermediate budgets was less than the national mean of $157.

To find a 90% confidence interval for the median (or mean when the distribution is symmetric) proceed as follows.

Menu commands:

1 Choose Stat ➤ Nonparametrics ➤ 1-Sample Wilcoxon...
2 Enter `COST` in the Variables: text box
3 Select the Confidence interval option button
4 Press Tab, and type 90 in the Level: text box
5 Select OK

Session commands:

- Type `WINTERVAL 90 'COST'` at the `MTB >` prompt and press Enter

Printout 9.5 shows the resulting output.

PRINTOUT 9.5
Minitab output for WINTERVAL

```
                   Estimated    Achieved
              N       Median  Confidence  Confidence Interval
COST         10        149.5        89.7  (   143.0,    155.5)
```

From the printout you see that an upper one-sided 95% confidence bound on the median is 155.5. Since the hypothesized value of 157 falls OUTSIDE this interval, the null hypothesis should be rejected as previously discovered. ■

PROBLEM 9.3 The suggested retail price for a certain brand of color TV is $599. Management suspects that the true mean price, μ, is *lower*. A random sample of eight stores revealed that the TV was offered for the prices shown in following table.

$565	505	558	642
583	617	645	509

Do the data confirm the suspicions of management? Apply WTEST to perform the required hypothesis test at the 5% significance level after making a judgment about symmetry. Use a column named TVPRICE for storing the data. Compute a 90% confidence interval for the true mean price. ◇

LESSON 10

Inferences for Two Population Means

GENERAL OBJECTIVE

In Lessons 8 and 9 we examined the Minitab commands required to find confidence intervals and perform hypothesis tests for one population mean, μ. Frequently, however, inferential statistics is used to compare the means of *two* populations. This lesson shows how Minitab can be employed to make inferences concerning two populations.

LESSON OUTLINE

10.1 Inferences for two population means (σs equal)

E-10.2:I-10.2

In the exercises in Section 10.1 in *IS/6e* and *ES/5e* two-sample z-procedures are presented for inference concerning two means when the populations are normal and both population standard deviations, σ_1 and σ_2, are known. However, as discussed there, such circumstances are rare. It has always been standard practice when the sample sizes are very large, to apply these procedures substituting sample estimates of the two population standard deviations and treat the results as approximate.

When the population standard deviations are unknown, there are two methods that apply as long as the populations being sampled are *normally distributed.* One requires that the populations under consideration have equal standard deviations and the other does not. Minitab has commands to handle both of these cases.

Procedure 10.1 in *IS/6e* and *ES/5e* gives a step-by-step method for performing a hypothesis test to compare the means of two normally distributed populations with equal standard deviations or large sample sizes, using independent samples. Procedure 10.2 in *IS/6e* and *ES/5e* determines confidence intervals for the difference of means in these same circumstances. Minitab has a command for carrying out both of these procedures directly. Since it is assumed that the two standard deviations are equal, though unknown, the data from both samples are combined to estimate this common standard deviation. Hence, the procedures are called *pooled.* The next example illustrates the use of this command.

EXAMPLE 10.1 *Illustrates the pooled t-test and pooled t-interval*

The annual salaries of faculty in public institutions is to be compared to the annual salaries of faculty in private institutions. A random sample of 30 salaries from public institutions and a random sample of 35 salaries from private institutions is taken. The salaries are displayed in Table 10.1 (WeissStats CD:Tb10-03).

TABLE 10.1 Annual salaries ($1000) for 30 faculty members in public institutions and 35 faculty members in private institutions

Public institutions						Private institutions						
34.2	63.6	24.4	79.4	33.8	88.2	92.9	102.2	51.5	77.6	71.1	59.3	71.0
90.0	56.8	56.0	42.4	40.2	44.6	52.0	102.2	51.5	77.6	71.1	59.3	71.0
100.4	41.4	58.2	81.8	51.2	64.4	63.1	53.8	45.2	78.3	67.6	27.2	92.6
24.6	35.0	76.8	29.2	41.2	74.0	118.5	101.0	76.0	66.3	52.4	81.2	56.0
107.4	54.2	84.2	15.8	60.2	71.0	37.7	68.6	56.1	31.3	47.2	24.8	62.3

At the 5% significance level, does it appear that there is a difference in mean salaries for faculty teaching in public and private institutions? Use Minitab to make the decision and assume that the standard deviations are equal.

SOLUTION Let μ_1 and μ_2 denote the actual mean salary for faculty in public and private institutions, respectively. Then the problem is to perform the hypothesis test

$$H_0: \mu_1 = \mu_2 \text{ (mean salaries are the same)}$$
$$H_a: \mu_1 \neq \mu_2 \text{ (mean salaries are different)}$$

at the 5% level of significance. Note that the test is two-tailed since there is a not-equal sign ($\neq$) in the alternative hypothesis.

By assumption, the samples are independent and the sample sizes are both large. Hence we are entitled to use Procedure 10.1. Enter the two sets of sample data from Table 10.1 into columns named PUBL and PRIV in an active worksheet. Proceed as follows.

Menu commands:

1 Choose Stat ➤ Basic Statistics ➤ 2-Sample t...
2 Select the Samples in different columns radio button
3 Select the First: text box and enter PUBL, select the Second: text box and enter PRIV
4 Be certain the Assume equal variances check box is checked
5 Click the [Options...] button
6 Select the Confidence level text box and type 95
7 Select the Test mean: text box and type 0
8 Click the [▾] button in the Alternative: text box and select not equal
9 Select [OK] twice

Session commands:

- Type TWOSAMPLE 95 'PUBL' 'PRIV'; at the MTB > prompt and press [Enter]
- Type POOLED. at the SUBC> prompt and press [Enter]

Printout 10.1 displays the results.

PRINTOUT 10.1
Minitab output for the pooled two-sample *t*-test

```
Two-sample T for PUBL vs PRIV

       N      Mean      StDev   SE Mean
PUBL  30      57.5       24.0       4.4
PRIV  35      66.4       22.3       3.8

Difference = mu PUBL - mu PRIV
Estimate for difference:  -8.91
95% CI for difference: (-20.38, 2.55)
T-Test of difference = 0 (vs not =): T-Value = -1.55  P-Value = 0.125  DF = 63
Both use Pooled StDev = 23.1
```

The first line of the output describes the Minitab output. The next three lines give the sample size, sample mean, sample standard deviation, and estimated standard error of the mean for each of the two samples, much as in the DESCRIBE command. Just below the summary of data, the 95% confidence interval for the difference of the two means is (-20.38, 2.55). The next-to-last line of the output gives a statement of the null and alternative hypotheses, followed by the value of the test statistic (`T= -1.55`), the P-value (`P=0.125`), and the degrees of freedom (`DF= 63`). On the last line, we find the value of the pooled sample standard deviation, s_p, listed as `Pooled Stdev = 23.1`.

Observe that the computed P-value of 0.125 is greater than the level of significance $\alpha = .05$. Using the `INVCDF 0.975;` command followed by the `t 63.` subcommand in Minitab, you may verify that the critical values for a classical test are ± 1.9983 and these limits cover the value of the test statistic, -1.55. Finally, 0 is included in the 95% confidence interval. By all three methods of testing, you should NOT reject the null hypothesis. In other words, the data do not provide sufficient evidence that there is a difference between the mean salaries of faculty teaching in public and private institutions. ■

PROBLEM 10.1 In a packing plant, a machine packs cartons with jars. A salesperson claims that the machine she is selling will pack faster. To test that claim, the time it takes each machine to pack 10 cartons is recorded. The results, in seconds, are in the table that follows.

New machine		Present machine	
42.0	41.0	42.7	43.6
41.3	41.8	43.8	43.3
42.4	42.8	42.5	43.5
43.2	42.3	43.1	41.7
41.8	42.7	44.0	44.1

Do the data suggest that the new machine packs *faster,* on the average? Use $\alpha = 0.05$. [Assume the packing times for both machines are normally distributed and that the standard deviations of the packing times for both machines are the same.] Employ Minitab to perform the appropriate hypothesis test. Use `NEW` as the name for the New machine data and `PRESENT` for the name of the Present machine data. ◇

10.2 Inferences for two population means (σs not equal)

E-10.3:I-10.3

Section 10.3 in *IS/6e* and *ES/5e* discusses inference to compare the means of two normally distributed populations using independent samples without making any assumptions about the relationship between the standard deviations of the two populations. The basis for the inference is the following key fact.

KEY FACT 10.1

Suppose that independent random samples of sizes n_1 and n_2 are to be taken from two normally distributed populations with means μ_1 and μ_2, respectively. Then the random variable

$$t = \frac{(\overline{x}_1 - \overline{x}_2) - (\mu_1 - \mu_2)}{\sqrt{(s_1^2/n_1) + (s_2^2/n_2)}}$$

has approximately the t-distribution with degrees of freedom given by

$$\Delta = \frac{\left[(s_1^2/n_1) + (s_2^2/n_2)\right]^2}{\dfrac{(s_1^2/n_1)^2}{n_1 - 1} + \dfrac{(s_2^2/n_2)^2}{n_2 - 1}}$$

rounded down to the nearest integer.

Thus the test statistic for a hypothesis test with null hypothesis $H_0 : \mu_1 = \mu_2$, is

$$t = \frac{(\overline{x}_1 - \overline{x}_2)}{\sqrt{(s_1^2/n_1) + (s_2^2/n_2)}},$$

and the endpoints of a $(1 - \alpha)$-level confidence interval for $\mu_1 - \mu_2$ are

$$(\overline{x}_1 - \overline{x}_2) \pm t_{\alpha/2} \cdot \sqrt{(s_1^2/n_1) + (s_2^2/n_2)}.$$

Procedure 10.3 in *IS/6e* and *ES/5e* gives a step-by-step method of carrying out a test under these circumstances. We will demonstrate this application in the next example using Minitab.

EXAMPLE 10.2 *Illustrates the nonpooled two-sample t-procedure*

A group of neurosurgeons wanted to see whether a dynamic system (Z-plate) reduced the operative time relative to a static system (ALPS plate). R. Jacobowitz, Ph.D., an ASU professor, along with G. Vishteh, M.D., and other neurosurgeons, obtained the following data (WeissStats CD:Tb10-07) on operative times, in minutes.

TABLE 10.2
Operative times, in minutes, for dynamic and static systems

Dynamic							Static		
370	360	510	445	295	315	315	430	445	455
345	450	505	335	280	325	500	455	490	535

At the 1% level of significance, do the data provide sufficient evidence to conclude that the mean operative time is less with the dynamic system than with the static system?

SOLUTION Let μ_1 and μ_2 denote the actual mean time for dynamic and static conditions, respectively. Then the problem is to perform the hypothesis test

$$H_0: \mu_1 = \mu_2 \text{ (mean dynamic time is the same as mean static time)}$$
$$H_a: \mu_1 < \mu_2 \text{ (mean dynamic time is less than mean static time)}$$

at the 5% level of significance. Note that the test is left-tailed since there is a less-than sign ($<$) in the alternative hypothesis.

Copy the data into columns named DYNA and STAT. Then proceed as follows.

Menu commands:

1 Choose Stat ➤ Basic Statistics ➤ 2-Sample t...
2 Select the Samples in different columns radio button
3 Select the First: text box and enter DYNA, select the Second: text box and enter STAT
4 Be certain the Assume equal variances check box is NOT checked
5 Click the [Options...] button
6 Select the Confidence level text box and type 99
7 Select the Test mean: text box and type 0
8 Click the [▾] button in the Alternative: text box and select less than
9 Select [OK] twice

Session commands:

- Type TWOSAMPLE 99 'DYNA' 'STAT'; at the MTB > prompt and press [Enter↵]
- Type ALTERNATIVE -1. at the SUBC> prompt and press [Enter↵]

Printout 10.2 displays the results.

PRINTOUT 10.2
Minitab output for the TWOSAMPLE command

```
Two-sample T for DYNA vs STAT

        N      Mean     StDev   SE Mean
DYNA   14     394.6      84.7        23
STAT    6     468.3      38.2        16

Difference = mu DYNA - mu STAT
Estimate for difference:  -73.7
99% upper bound for difference: -3.1
T-Test of difference = 0 (vs <): T-Value = -2.68  P-Value = 0.008  DF = 17
```

The output provides the same information as the pooled test. Just below the summary of data, the hypothesized difference is listed as mu DYNA - mu STAT, followed by a point estimate of the difference, −73.7.

The 99% upper bound is given so that the upper one-sided 99% confidence interval for the difference of the two means is $(-\infty, -3.1)$. The next-to-last line of the output gives a statement of the null and alternative hypotheses, followed by the value of the test statistic (`T= -2.68`), the P-value (`P=0.008`), and the degrees of freedom (`DF= 17`).

Observe that the computed P-value of 0.008 is less than the level of significance $\alpha = .01$. Using the `INVCDF 0.01;` command followed by the `t 17.` subcommand Minitab, you may verify that the critical value for a classical test is -2.5669, yielding a rejection region $(-\infty, -2.5669)$. The value of the test statistic, -2.68 is inside this region. Finally, 0 is NOT included in the 99% one-sided confidence interval. By all three methods of testing, you should reject the null hypothesis. In other words, at the 1% level of significance, the data provide sufficient evidence to conclude that the mean operative time is less with the dynamic system than with the static system, contrary to the statement in the null hypothesis.

You might observe that in the Minitab output for Printout 10.2 in *IS/6e* and *ES/5e* a 99% two-sided confidence interval is given as (-153.4, 6.0). Computing this confidence interval may be of interest in itself but you must be careful about relating it to the hypothesis test since the alternative to the hypothesis is one-sided. Indeed, this very example illustrates why care needs to be exercised. If you were to decide on the viability of the null hypothesis based on whether or not you find 0 in the *two-sided* 99% confidence interval, you would FAIL to reject the null hypothesis since 0 lies in the interval. This is contrary to the conclusion drawn by the classical test or the use of the correct 99% one-sided confidence interval! ■

PROBLEM 10.2

The owner of a chain of car washes needs to decide between two brands of hot waxes. One of the brands, Sureglow, costs less than the other brand, Mirror-Sheen. Therefore, unless there is strong evidence that the second brand outlasts the first, the owner will purchase the first brand. With the cooperation of several local automobile dealers, 30 cars are randomly selected to take part in the test. Fifteen of the 30 cars are waxed with Sureglow and 15 with Mirror-Sheen. The cars are then exposed to the same environmental conditions. The test results below show the effectiveness times, in days.

Sureglow			Mirror-Sheen		
87	90	88	92	92	91
93	90	92	91	92	91
91	89	93	93	93	92
88	87	89	92	93	94
91	91	90	93	94	91

At the 1% significance level, does Mirror-Sheen seem to have a longer effectiveness time, on the average, than Sureglow? [Assume that the effectiveness times for both waxes are normally distributed.] Name the columns where you store the data `SURE` and `MIRROR`. ◇

10.3 The Mann–Whitney test

I-10.4

We will now examine a nonparametric test for comparing the medians (or means) of *two* populations using independent samples. The test is called the *Mann-Whitney test* and is also referred to as the *Wilcoxon rank-sum test.* An assumption for use of the Mann-Whitney test is that the two populations under consideration have the same shape. The corresponding Minitab program is called MANNWHITNEY. Example 10.3 shows how MANNWHITNEY is applied.

EXAMPLE 10.3 Illustrates the MANNWHITNEY command

A nationwide shipping firm purchased a new computer system and decided to hire a technical writer to write a short self-study manual for that purpose. The manual took very little time for some people and quite a bit of time for others. Someone suggested that this could have happened because some employees had previous experience with computers and others did not. To test this suggestion, the company decided to time a sample of 15 employees with similar educational levels but different computer backgrounds. The times required for the 15 employees to learn to use the computer are displayed in Table 10.3 (WeissStats CD:Tb10-09).

TABLE 10.3 Times, in minutes, seven employees without computer experience and eight employees with computer experience.

Without experience	With experience
139	142
118	109
164	130
151	107
182	155
150	88
134	95
	104

At the 5% significance level, do the data provide evidence that the mean learning time for employees without computer experience is *greater* than that for employees with computer experience? Compute a 90% confidence interval for the difference in medians for use as a test.

SOLUTION Let μ_1 and μ_2 denote the mean learning times for employees without computer experience and with computer experience, respectively. Then the null and alternative hypotheses are

$$H_0\text{: } \mu_1 = \mu_2 \text{(mean time for inexperienced employees is not greater)}$$
$$H_a\text{: } \mu_1 > \mu_2 \text{(mean time for inexperienced employees is greater)}$$

Note that the test is right-tailed because there is a greater-than sign (>) in the alternative hypothesis.

To apply MANNWHITNEY, first enter the two data sets from Table 10.3 into columns named WOUT and WITH in the usual manner. Then proceed as follows.

Menu commands:

1 Choose Stat ➤ Nonparametrics ➤ Mann-Whitney...

2 Enter WOUT in the First Sample: dialog box and enter WITH in the Second Sample: dialog box

3 Click the ▾ button in the Alternative box and select greater than as the alternative

4 Select the Confidence level box and type 90

5 Select [OK]

Session commands:

- Type MANN 90 'WOUT' 'WITH'; at the MTB > prompt and press [Enter↵]
- Type ALTE 1. at the SUBC> prompt and press [Enter↵]

See Printout 10.3 for the results. The output first gives the sample sizes and sample medians. Then it provides a point estimate for the difference, $\eta_1 - \eta_2 = 31.5$, between the population medians (or means).

PRINTOUT 10.3 Minitab output for the MANNWHITNEY command

```
Mann-Whitney Test and CI: WOUT, WITH

WOUT        N =   7     Median =      140.00
WITH        N =   8     Median =      108.00
Point estimate for ETA1-ETA2 is        31.50
90.7 Percent CI for ETA1-ETA2 is (8.99,52.00)
W = 74.0
Test of ETA1 = ETA2  vs  ETA1 > ETA2 is significant at 0.0214
```

Next comes a confidence interval for $\eta_1 - \eta_2$ (or $\mu_1 - \mu_2$) with a confidence level as close to 90% as possible, 90.7% in this case. Thus, a lower one-sided 95% confidence interval (consistent with an upper-tailed test) covers any value from about 9 on up. Since 0 is not in the interval, we would be inclined to reject the hypothesis of equality in favor of the alternative. Indeed, we would be inclined to think that the true difference is at least 9 minutes. In the next-to-the-last line, we find the value of the test statistic,

$$M = \text{sum of the ranks for sample data from Population 1}$$

(Minitab uses the letter W instead of M). Thus $M = 74$.

Finally, the last line of the output gives a statement of the null and alternative hypotheses and the P-value. Since the P-value of 0.0214 is less than the designated significance level of $\alpha = 0.05$, we reject H_0. In other words, the data indicate that the median learning time for employees without computer experience is greater than that for employees with computer experience. This is consistent with the confidence interval results. ■

When there are ties in the sample data, ranks are assigned by averaging ranks. That is, if two or more data values are tied, each is assigned the average of the ranks they would have had if there were no ties. The next example gives a further illustration of the Mann Whitney test and also provides a case in which ties occur. We shall see how Minitab adjusts for the ties.

EXAMPLE 10.4 Illustrates MANNWHITNEY with ties

The U.S. National Center for Health Statistics collects information on the ages of Americans at the time of their first marriage. Data are published in *Vital Statistics of the United States.* Independent random samples of 10 married males and 10 married females yielded the ages at first marriage shown in Table 10.5. Can we conclude that the mean ages at the time of first marriage are different for men and women? Perform a Mann-Whitney test at the 5% significance level using Minitab.

SOLUTION Let μ_1 and μ_2 denote the mean ages at first marriage for males and females, respectively. We wish to use the Mann-Whitney test to perform the hypothesis test:

$$H_0\colon \mu_1 = \mu_2 \text{(mean ages are the same)}$$
$$H_a\colon \mu_1 \neq \mu_2 \text{(mean ages are different)}$$

at the 5% significance level. Note that the hypothesis test is two-tailed since there is a not-equal sign ($\neq$) in the alternative hypothesis.

TABLE 10.4
Ages at first marriage

Male	Female
27	20
23	19
19	28
36	23
33	23
22	27
27	22
21	34
22	31
41	23

To apply MANNWHITNEY, first enter the sample data from Table 10.4 into columns named `MALE` and `FEMALE` in the usual manner. Proceeding as in the last example (with confidence 95%), the results of the test may be seen in Printout 10.4.

PRINTOUT 10.4 Minitab output for the MANNWHITNEY command

```
Mann-Whitney Test and CI: MALE, FEMALE

MALE        N =  10     Median =        25.00
FEMALE      N =  10     Median =        23.00
Point estimate for ETA1-ETA2 is          0.50
95.5 Percent CI for ETA1-ETA2 is (-4.00,8.00)
W = 109.0
Test of ETA1 = ETA2  vs  ETA1 not = ETA2 is significant at 0.7913
The test is significant at 0.7899 (adjusted for ties)

Cannot reject at alpha = 0.05
```

As we see from Printout 10.4, the output first gives the sample sizes and the sample medians for the two samples. Then it provides a point estimate and confidence interval for the difference, $\eta_1 - \eta_2$, between the population medians. The confidence level is chosen to be as close to 95% as possible. Since 0 is included in the confidence interval, we should not reject the null hypothesis.

Next we find the value of the test statistic, listed in Minitab as `W = 109.0`. However, recall that this statistic is labeled M in *IS/6e* and *ES/5e*. In terms of the notation used in your book then, we may state:

$$M = \text{sum of the ranks for sample data from Population 1,}$$

for a Mann–Whitney test.

Below the value of the test statistic is a statement of the null and alternative hypotheses and an approximate P-value:

```
Test of ETA1 = ETA2 vs ETA1 not = ETA2 is significant at 0.7913.
```

Minitab employs a normal approximation with a continuity correction factor to obtain the approximate P-value. If there are ties in the data, as there are in this example, another approximate P-value is printed:

```
The test is significant at 0.7899 (adjusted for ties).
```

This second approximation is usually closer to the actual P-value than the first approximation.

Since the (approximate) P-value is 0.7899, which exceeds the designated significance level of $\alpha = 0.05$, we do not reject H_0. In other words, the data do not provide sufficient evidence to conclude that there is a difference in the mean ages at first marriage for males and females. ■

PROBLEM 10.3 The U.S. Bureau of Labor Statistics gathers data on weekly earnings of full-time wage and salary workers, by sex. Independent random samples of male and female workers gave the earnings (in dollars) listed in the following table.

Male	Female
141	248
762	277
392	284
316	350
425	230
339	437
470	156
678	260
	316
	435

At the 5% significance level, do the data indicate that the mean weekly earnings of male full-time wage and salary workers is *greater* than that for females? Use MANNWHITNEY in Minitab to carry out the hypothesis test, naming the columns `MEN` and `WOMEN`. Also compute a 90% confidence interval and use the result to analyze the hypothesis in terms of a one-sided 95% confidence interval. ◇

10.4 Inferences for two population means using paired samples

E-10.4:I-10.5

Up to now we have considered inferences for two population means when the samples are *independent*. However, it is often more suitable to use *paired samples*. With paired samples, each item in the sample consists of a pair of numbers, one from each of the two populations.

By pairing the samples, when it is appropriate, we can remove extraneous sources of variation. This, in turn, makes it more likely that we will detect a difference between the two population means, when such a difference exists.

We can think of the paired differences of the pairs sampled as a random sample from the population of all possible paired differences. If that population is *normally distributed,* then we can apply the following key fact for deriving tests and confidence intervals.

KEY FACT 10.2

Suppose that a random sample of n pairs is to be taken from populations with means μ_1 and μ_2. Further suppose that the population of all paired differences, represented by the random variable d, is normally distributed. Then the random variable

$$t = \frac{\overline{d} - (\mu_1 - \mu_2)}{s_d/\sqrt{n}}$$

has the t-distribution with df $= n - 1$.

Note: We will use the phrase *normal differences* as an abbreviation of "the population of paired differences is normally distributed."

Procedure 10.6 in *IS/6e* (Procedure 10.5 in *ES/5e*) gives a step-by-step method for performing a hypothesis test to compare the means of two normally distributed populations when the samples are paired. Procedure 10.7 in *IS/6e* (Procedure 10.6 in *ES/5e*) determines confidence intervals for the difference of means in these same circumstances. Minitab has a command called PAIRED for carrying out both of these procedures directly using various subcommands or menu choices to achieve the desired results. The next example illustrates the use of this command.

EXAMPLE 10.5 *Illustrates the use of PAIRED to perform a paired-difference test*

A major oil company has developed a new gasoline additive that is supposed to increase mileage. To test that hypothesis, 10 cars are randomly selected. Each car sampled is driven both with and without the additive. The resulting gas mileages, in miles per gallon, are displayed in the second and third columns of Table 10.6 (WeissStats CD:Tb10-12) that we present next.

TABLE 10.5 Gas mileages, with and without additive, for 10 randomly selected cars

Car	With additive x_1	Without additive x_2
1	25.7	24.9
2	20.0	18.8
3	28.4	27.7
4	13.7	13.0
5	18.8	17.8
6	12.5	11.3
7	28.4	27.8
8	8.1	8.2
9	23.1	23.1
10	10.4	9.9

Do the data suggest that, on the average, the gasoline additive improves mileage? Use Minitab to perform the appropriate hypothesis test at the 5% level of significance. Note that the samples are *paired* since each mileage is associated with the same car. [Assume that differences in mileage are normally distributed.]

SOLUTION Let μ_1 denote the mean gas mileage of all cars when the additive is used and μ_2 denote the mean gas mileage of all cars when the additive is not used. Then we want to perform the hypothesis test

$$H_0: \mu_1 = \mu_2 \text{ (mean mileage with additive is not greater)}$$
$$H_a: \mu_1 > \mu_2 \text{ (mean mileage with additive is greater)}$$

at the 5% significance level.

First, enter the paired samples into columns named ADD and NOADD in the usual manner. Then proceed as follows.

Menu commands:

1 Choose Stat ➤ Basic Statistics ➤ Paired t...

2 Enter ADD in the First sample: text box and enter NOADD in the Second sample: text box

3 Click the [Options...] button

4 Type 95 in the Confidence level: text box if necessary

5 Select the Test mean: text box and type 0 if necessary

6 Click the [▾] button in the Alternative: text box and select greater than from the drop down list

7 Select [OK] twice

Session commands:

- Type `PAIRED 'ADD' 'NOADD';` at the MTB > prompt and press [Enter↵]
- Type `TEST 0;` at the SUBC> prompt and press [Enter↵]
- Type `ALTERNATIVE 1.` at the SUBC> prompt and press [Enter↵]

Printout 10.5 displays the resulting output.

PRINTOUT 10.5 Minitab output for the PAIRED procedure applied to the data.

```
Paired T for ADD - NOADD

                 N      Mean     StDev   SE Mean
ADD             10     18.91      7.47      2.36
NOADD           10     18.25      7.42      2.35
Difference      10     0.660     0.443     0.140

95% lower bound for mean difference: 0.403
T-Test of mean difference = 0 (vs > 0): T-Value = 4.71  P-Value = 0.001
```

We see from Printout 10.5 that the P-value is 0.001. Since the P-value is less than the designated significance level of $\alpha = 0.05$, we reject H_0. Also, we note that the lower one-sided 95% confidence interval is $(0.403, \infty)$ and the value 0 is NOT in the confidence interval, so that we would reject H_0 on that account.

Thus, the data provide sufficient evidence to conclude that the mean gas mileage of all cars when the additive is used is greater than the mean gas mileage of all cars when the additive is not used. In other words, it appears that the additive is effective in increasing gas mileage. ■

PROBLEM 10.4

An agronomist has developed a new variety of wheat that she feels will return a greater yield. For an initial test, she selects 10 wheat farms at random. On each farm, two adjacent one-acre plots are chosen. A widely-used current variety is planted on one of the one-acre plots and the new variety on the other. The yields, in bushels, are shown in the following table.

Farm	Current variety x_1	New variety x_2
1	34	37
2	40	35
3	34	34
4	33	34
5	42	44
6	35	39
7	34	33
8	42	44
9	41	43
10	39	40

Assume that changes in crop yields due to wheat varieties are normally distributed.

Do the data provide evidence, at the 5% significance level, that the new variety of wheat gives a better yield, on the average, than the current variety? Use Minitab to find the answer. Use `CURRENTV` and `NEWV` as names for the two columns of data and `DIFF` for the difference. Compute a 90% confidence interval. ◇

10.5 The Paired Wilcoxon signed-rank test **I-10.6**

If the population of all paired differences is symmetric (but not necessarily normal), then we can use the Wilcoxon signed-rank test to perform a hypothesis test to compare the means of two populations. Essentially, we may apply Procedure 9.4 of *IS/6e* , with null hypothesis $H_0 : \mu = 0$, to the sample of paired differences. The steps are summarized in Procedure 10.8 of *IS/6e* . With Minitab, we need only apply the `WTEST` command to achieve the same results.

EXAMPLE 10.6 *Illustrates Procedure 10.8 using Minitab*

Let us return to the gas-mileage problem of Example 10.5. Using those data, we apply the Wilcoxon paired-sample signed-rank test to perform a test of the same hypothesis and alternative.

SOLUTION We assume that the data from the example are still stored in the two columns `ADD` and `NOADD`. First we will compute the paired differences and store them in a column named `DIFFER`.

Menu commands:

1 Choose Calc ➤ Calculator...

2 Type DIFFER in the Store results in variable: text box

3 Specify ADD−NOADD in the Expression: text box

4 Select [OK]

Session commands:

- Name an empty column DIFFER if necessary
- Type `LET 'DIFFER'='ADD'-'NOADD'` at the MTB > prompt and press Enter

Now that the paired difference are accessible, proceed further to implement the Wilcoxon test as follows.

Menu commands:

1 Choose Stat ➢ Nonparametrics ➢ 1-Sample Wilcoxon...
2 Enter DIFFER in the Variables: text box
3 Press the Test median: option button and type 0 in the input box if necessary
4 Click the ▾ button in the Alternative: text box and select greater than from the drop-down list
5 Select OK

Session commands:

- Type `WTEST 0 'DIFFER';` at the MTB > prompt and press Enter
- Type `ALTE 1.` at the SUBC> prompt and press Enter

The results are shown in Printout 10.6.

PRINTOUT 10.6

```
Test of median = 0.000000 versus median  >  0.000000

                N for   Wilcoxon              Estimated
           N    Test   Statistic        P        Median
DIFFER    10       9        44.0    0.006        0.7000
```

As you see from the small size of the P-value, you would reject H_0 as in the t procedure.

To find a 90% confidence interval proceed as follows.

Menu commands:

1 Choose Stat ➢ Nonparametrics ➢ 1-Sample Wilcoxon...
2 Enter DIFFER in the Variables: text box
3 Press the Confidence interval option button
4 Press Tab, and type 90 in the Level: text box
5 Select OK

Session commands:

- Type `WINTERVAL 90 'DIFFER'` at the MTB > prompt and press Enter

Printout 10.7 shows the resulting output.

PRINTOUT 10.7
Minitab output
for WINTERVAL

```
                 Estimated    Achieved
              N     Median  Confidence  Confidence Interval
DIFFER       10      0.700        89.7  (   0.350,     0.950)
```

From the printout you see that a lower one-sided 95% confidence interval on the difference is $(0.350, \infty)$. Since the hypothesized value of 0 falls *outside* this interval, the null hypothesis should be rejected as previously discovered. ■

PROBLEM 10.5

For the circumstances of Problem 10.4, test using the Wilcoxon paired-sample signed-rank test and compare the results. ◇

LESSON 11

Inferences for Population Standard Deviations

GENERAL OBJECTIVE

In previous lessons, we have examined the Minitab commands that are used for statistical inferences concerning population means. Now we will learn how Minitab can be applied to perform tests and compute confidence intervals for standard deviations.

LESSON OUTLINE

11.1 Inferences for one population standard deviation

I-11.1

Recall that the *standard deviation,* σ, of a population is a measure of dispersion or variability of the population values. Populations with large standard deviations have a large spread, while those with small standard deviations have a small spread.

There are many circumstances where the standard deviation of a population is unknown, but where it is nevertheless important to have information about its value. In most cases, it is impossible to determine the value of the population standard deviation exactly. Consequently, we employ inferential statistics to obtain the required information.

In this lesson, we will learn how to perform hypothesis tests and construct confidence intervals for the standard deviation of a normally distributed random variable. The inferential procedures discussed in this lesson all rely on a class of continuous probability distributions called chi-square distributions. We pause to study that family of distributions.

THE χ^2-DISTRIBUTION

Probabilities for a random variable having a chi-square distribution are equal to areas under a curve–a χ^2 `(chi-square) curve`(see Figure 11.1).

FIGURE 11.1
A typical χ^2-curve

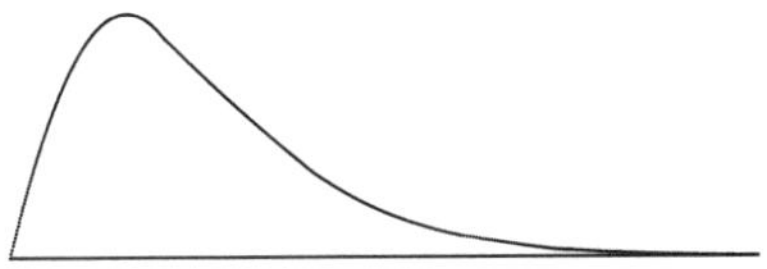

To perform a hypothesis test or obtain a confidence interval that is based on a chi-square distribution, we will need to be able to find the χ^2-value(s) corresponding to a specified area under a χ^2-curve. Recall that the symbol χ^2_α is used to represent the χ^2-value with area α to its right under a χ^2-curve. This quantity can be found using Table V in *IS/6e* and *ES/5e* or, alternatively, with the aid of Minitab. In addition, we need to compute the area under a χ^2 (chi-square) curve in order to compute P-values for test statistics.

Up to now the distributions we have been dealing with for statistical inference have been symmetric. That is, the area to the right of the mean is a mirror image of the area to the left. This is not true with the χ^2 family of distributions (see Figure 11.1). Like the t family, each χ^2 distribution is distinguished by a parameter known as degrees of freedom, a positive whole number that we will have to supply to Minitab in order to produce a desired output.

Minitab has a subcommand called CHISQUARE which, with a specification of the degrees of freedom, may be used with both the CDF and the INVCDF commands.

In conjunction with the INVCDF command, this subcommand may be used to compute the χ^2-value with a specified area to its *left*. From this, it is a simple matter to obtain the χ^2-value(s) corresponding to any specified area under an arbitrary χ^2-curve. We present the details in the example to follow.

EXAMPLE 11.1 *Illustrates how to find the χ^2-value for a specified area using Minitab*

For a χ^2-curve with 20 degrees of freedom, find $\chi^2_{0.05}$; that is, find the χ^2-value with area 0.05 to its right. See Figure 11.2.

FIGURE 11.2
The χ^2-curve with 20 degrees of freedom

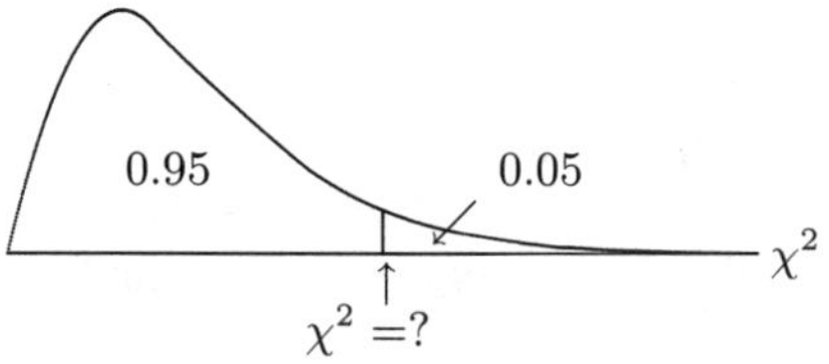

SOLUTION To find the χ^2-value with area 0.05 to its right, we observe that the area to the left must be 0.95. Proceed as follows.

Menu commands:

1 Choose Calc ➤ Probability Distributions ➤ Chi-square...
2 Select the Inverse cumulative probability radio button
3 Select the Degrees of freedom: text box and type 20
4 Select the Input constant: radio button
5 Press Tab and type 0.95 in the text box
6 Select OK

Session commands:

- Type `INVCDF 0.95;` at the MTB > prompt and press Enter
- Type `CHISQUARE 20.` at the SUBC> prompt and press Enter

PRINTOUT 11.1
Finding a χ^2-value

```
Chi-Square with 20 DF

 P( X <= x)          x
    0.9500     31.4104
```

From the resulting printout, we see that $\chi^2_{0.05} = 31.4104$. ■

PROBLEM 11.1

For a χ^2-curve with 14 degrees of freedom, use Minitab to find
a) the χ^2-value with area 0.10 to its left.
b) the χ^2-value with area 0.10 to it right. ◇

To calculate P-values, we need to be able to find the area under a χ^2-curve for given values. Using the CDF command in Minitab, we can find such areas.

EXAMPLE 11.2 Illustrates finding the area under a χ^2-curve using Minitab

For a χ^2-curve with 20 degrees of freedom, find the area to the right of the value $\chi^2 = 23$. See Figure 11.3.

FIGURE 11.3
The χ^2-curve with 20 degrees of freedom

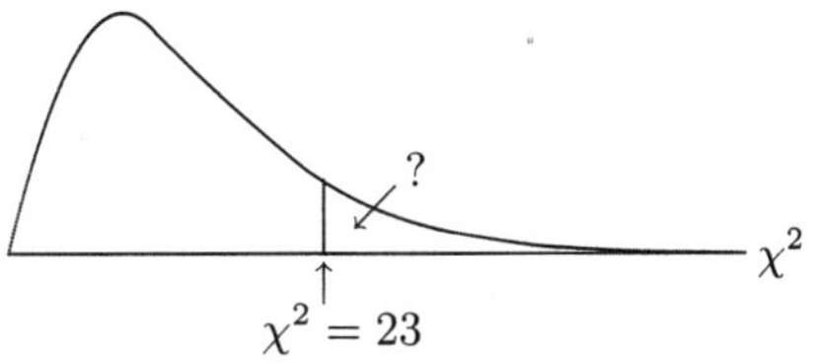

SOLUTION Recall that the CDF command in Minitab always produces the area to the *left* of any value that we specify. Hence we will need to find that area for the χ^2 value of 23 first. Proceed as follows.

Menu commands:

1 Choose Calc ➤ Probability Distributions ➤ Chi-square...
2 Select the Cumulative probability radio button
3 Select the Degrees of freedom: text box and type 20
4 Select the Input constant: radio button
5 Press [Tab] and type 23 in the text box
6 Select [OK]

Session commands:

- Type CDF 23; at the MTB > prompt and press [Enter]
- Type CHISQUARE 20. at the SUBC> prompt and press [Enter]

The result of either of these commands is displayed in Printout 11.2.

PRINTOUT 11.2
Finding a χ^2-value

```
Chi-Square with 20 DF

        x       P( X <= x)
  23.0000          0.7112
```

From the printout, we see that $P(\chi^2 \leq 23) = 0.7112$. Hence the area to the right of $\chi^2 = 23$ is approximately 0.29. ■

PROBLEM 11.2 For a χ^2-curve with 14 degrees of freedom, use Minitab to find
a) the area to the right of $\chi^2 = 13$.
b) the area between $\chi^2 = 7$ and $\chi^2 = 13$. ◇

HYPOTHESIS TESTS FOR A POPULATION STANDARD DEVIATION

We will now discuss how Minitab can be used to help carry out the inferential procedures for a population standard deviation that are presented in Section 11.1 of *IS/6e* . We begin with hypothesis tests and take up the matter of confidence intervals later.

Minitab does not have a specific program to perform a hypothesis test for a population standard deviation. However, we can use Minitab primitive commands in the session window to do the calculations required in implementing Procedure 11.1 in *IS/6e* . Consider the following example where the steps are repeated.

EXAMPLE 11.3 *Illustrates how Minitab can be used in conjunction with Procedure 12.3*

A hardware manufacturer produces 10-millimeter bolts. The manufacturer needs to decide whether the standard deviation, σ, of bolt diameters is *less than* 0.09 mm. He randomly samples 12 bolts and obtains their diameters. The results are shown in Table 11.1 (WeissStats CD:Tb11-01).

TABLE 11.1 Diameters, in mm, of 12 randomly selected bolts

10.05	10.00	10.02	9.97
10.07	10.03	9.98	10.10
9.95	9.99	10.00	10.08

At the 5% significance level, do the data provide evidence that the standard deviation, σ, of bolt diameters is less than 0.09 mm? [Assume that the population of diameters of all bolts manufactured is normally distributed.] Use Minitab to compute the results.

SOLUTION Enter the data in a column named `DIAM` in the usual manner. We apply Procedure 11.1.

STEP 1 *State the null and alternative hypotheses.*

The null and alternative hypotheses are

$$H_0\colon \sigma = 0.09 \text{ mm (too much variation)}$$
$$H_a\colon \sigma < 0.09 \text{ mm (not too much variation)}$$

Note that the test is left-tailed since there is a less-than sign (<) in the alternative hypothesis.

STEP 2 *Decide on the significance level, α.*

The test is to be performed at the 5% significance level. Thus, $\alpha = 0.05$.

STEP 3 *The critical value for a left-tailed test is $\chi^2_{1-\alpha}$, with df = $n - 1$.*

We have $\alpha = 0.05$ and $n = 12$. Consequently, df = 12 − 1 = 11. Therefore, we need to determine $\chi^2_{1-0.05} = \chi^2_{0.95}$ for df = 11. Since $\chi^2_{0.95}$ is the χ^2-value with area 0.95 to its right, it is also the χ^2-value with area 0.05 to its left. This last χ^2-value can be obtained by applying the command INVCDF and the CHISQUARE subcommand, as explained in previously. See Printout 11.3 for the result.

PRINTOUT 11.3
Finding the χ^2-value

```
Chisquare with 11 d.f.

 P( X <= x)          x
    0.0500        4.5748
```

Printout 11.3 shows that, for df = 11, the χ^2-value with area 0.05 to its left is 4.5748. This, as we have noted above, is also the desired critical value; that is, $\chi^2_{0.95} = 4.5748$.

STEP 4 *Compute the value of the test statistic*

$$\chi^2 = \frac{n-1}{\sigma_0^2} s^2$$

First we have Minitab compute the sample standard deviation, s, of the data in Table 11.4. Enter the data into a column named DIAM. As a session command, type LET K2=STDEV('DIAM') . The sample standard deviation, s, of the bolt diameters in Table 11.4 is now stored in K2.

Since $n = 12$ and $\sigma_0 = 0.09$, we have Minitab determine the value of the test statistic by typing

```
LET K3=(12-1)*K2**2/0.09**2
```

followed by PRINT K3. See Printout 11.4 for the result.

PRINTOUT 11.4
Computing the test statistic with Minitab

```
MTB > PRINT K3
K3        2.98766
```

Printout 11.4 shows that the value of the test statistic is

$$\chi^2 = \frac{n-1}{\sigma_0^2} s^2 = 2.98766$$

STEP 5 *If the value of the test statistic falls in the rejection region, reject H_0; otherwise, do not reject H_0.*

From Step 4, the value of the test statistic is $\chi^2 = 2.98766$. Since this value lies to the left of the critical value $\chi^2_{0.95} = 4.5748$ and the test is left-tailed, we reject H_0. Alternatively, we can find the P-value of the test using Minitab as shown in Printout 11.5.

PRINTOUT 11.5
Finding the *P*-value with Minitab

```
MTB > CDF 2.98766;
SUBC> CHISQUARE 11.
Chisquare with 11 d.f.

         x        P( X <= x)
    2.9877          0.0091
```

From Printout 11.5, we see that the area to the left of the test statistic is 0.0091. Since the test is left-tailed, this is the P-value of the test. Accordingly, since we have $0.0091 < 0.05$, H_0 should be rejected at the 5% level of significance.

STEP 6 *State the conclusion in words.*

It appears that the standard deviation, σ, of the diameters of all bolts being produced is less than 0.09 mm. We offer no menu commands for this example. ■

With all of the steps involved in Procedure 11.1, this is a natural place to construct a macro since Minitab does not provide a separate routine for this case. Such a macro is included in the *WeissStats CD* for you. This is a local macro that accepts a single variable, namely, a column of data. To run this macro, we assume it is located in the Macro folder of your Minitab installation. If not, adjust the calls in the following commands accordingly.

Session commands:

1 Type `%1stdev 'DIAM'`

2 Type `Y` in response to the question, Do you want to perform a hypothesis test (Y/N)? and press [Enter↵]

3 Type `0.09` in response to Enter the null hypothesis standard deviation and press [Enter↵]

4 Type `-1` in response to Enter 0, 1, or -1, respectively, for a two-tailed, right-tailed, or left-tailed test and press [Enter↵]

5 Type `N` in response to the question, Do you want a confidence interval (Y/N)? and press [Enter↵]

Printout 11.6 on the following page shows the resulting output, confirming the tedious calculations made above.

PRINTOUT 11.6 Output of the %macro 1stdev

```
Test of sigma =      0.09000 vs sigma <      0.09000


 Row  Variable     n  StDev  Chi-Sq         P

   1       DIAM   12  0.047   2.988  0.009
```

PROBLEM 11.3 Every year thousands of college-bound high-school students take the *Scholastic Aptitude Test* (SAT). This test, provided by the Educational Testing Service of Princeton, New Jersey, measures the verbal and mathematical abilities of prospective college students. Student scores are reported on a scale that ranges from a low of 200 to a high of 800. This scale was introduced in 1941. At that time, the standard deviation of scores was 100 points. Suppose that a random sample of 25 verbal scores for this year gives the following results:

560	405	367	416	540
347	370	629	570	372
396	439	526	610	339
569	314	228	483	353
379	558	432	648	475

At the 5% significance level, do the data suggest that the standard deviation, σ, for this year's verbal scores is *different* from the 1941 standard deviation of 100? Use the %macro 1stdev to perform the hypothesis test. Store the data in a column named `VERBAL`. ◇

CONFIDENCE INTERVALS FOR A POPULATION STANDARD DEVIATION

We can also use Minitab to aid us in constructing a confidence interval for a population standard deviation, σ. In the next example, we will see how Minitab can be applied to help perform Procedure 11.2 of *IS/6e* .

EXAMPLE 11.4 *Illustrates how Minitab can be used in conjunction with Procedure 11.2*

Refer to Example 11.3. Use the sample data in Table 11.1 to find a 95% confidence interval for the standard deviation, σ, of the diameters of all bolts produced by the manufacturer. Employ Minitab in conjunction with Procedure 11.2 of *IS/6e* .

SOLUTION We repeat the steps of Procedure 11.2 on the following page.

STEP 1 *For a confidence level of 1 − α, find* $\chi^2_{1-\alpha/2}$ *and* $\chi^2_{\alpha/2}$ *for df = n − 1.*

We want a 95% confidence interval. This means $\alpha = 0.05$. Also, $n = 12$, which makes df = 12 − 1 = 11. Now,

$$\chi^2_{\alpha/2} = \chi^2_{0.05/2} = \chi^2_{0.025}$$

and

$$\chi^2_{1-\alpha/2} = \chi^2_{1-0.05/2} = \chi^2_{0.975}$$

We will use Minitab to find these two χ^2-values. The areas to the *left* of the two χ^2-values are, respectively, 0.975 and 0.025. Let us (1) store these area values in C10, (2) apply the INVCDF command and the CHISQUARE subcommand to obtain the desired χ^2-values, and (3) store these χ^2-values in C11. The details are shown in Printout 11.7.

PRINTOUT 11.7
Finding χ^2-values with Minitab

```
MTB > SET C10
DATA> .975 .025
DATA> END
MTB > INVCDF C10 C11;
SUBC> CHISQUARE 11.
MTB >
```

STEP 2 *The confidence interval for σ is*

$$\sqrt{\frac{n-1}{\chi^2_{\alpha/2}}}\cdot s \quad \textit{to} \quad \sqrt{\frac{n-1}{\chi^2_{1-\alpha/2}}}\cdot s$$

In Step 1 we stored $\chi^2_{\alpha/2}$ and $\chi^2_{1-\alpha/2}$ in C11. Furthermore, in Example 11.3, we stored in K2 the sample standard deviation, s, of the bolt diameters in Table 11.1. Recalling that $n = 12$, we can have Minitab obtain the endpoints of the confidence interval by typing

LET C12=SQRT((12-1)/C11)*K2

followed by PRINT C12. See Printout 11.8 for the results.

PRINTOUT 11.8
Finding the confidence endpoints with Minitab

```
MTB > LET C12=SQRT((12-1)/C11)*K2
MTB > PRINT C12
C12
  0.0332267   0.0796377
```

Printout 11.8 shows that a 95% confidence interval for σ is from 0.0332267 to 0.0796377. We can be 95% confident that the standard deviation, σ, of all bolt diameters is somewhere between 0.0332267 mm and 0.0796377 mm. ■

You might recall that the 1stdev macro prompted for a confidence interval at Step 5. You may use the macro to compute only a confidence interval as follows.

Menu commands:

1 Type %1stdev 'DIAM'

2 Type N in response to the question, Do you want to perform a hypothesis test (Y/N)? and press Enter↵

3 Type Y in response to the question, Do you want a confidence interval (Y/N)? and press Enter↵

4 Type 90 in response to Enter the confidence level, as a percentage and press Enter↵

Printout 11.9 shows the resulting output, confirming the tedious calculations made above.

PRINTOUT 11.9
Confidence interval using 1stdev

```
Row  Variable     n  StDev  Level     CI for sigma

  1  DIAMETER   12  0.047  90.0%  (0.035, 0.073)
```

The 90% confidence interval was appropriate because the test was one-sided. Since it was left-tailed, we see that the upper one-sided 95% confidence bound on σ is 0.073. Not surprisingly, we reject the null hypothesis since the hypothesized standard deviation falls outside the confidence interval, that is, $0.09 > 0.073$.

PROBLEM 11.4 Refer to Problem 11.3. Determine a 95% confidence interval for the standard deviation, σ, of this year's verbal SAT scores. Use Minitab in conjunction with Procedure 11.2 of *IS/6e* . ◇

11.2 Inferences for two population standard deviations

I-11.2

Tests and confidence intervals for one mean extended to the same considerations for two means. Hence one would expect an extension of the last section to the case of two standard deviations. In the case of two means, the theoretical structure that yielded the tests was based on sums or differences of random variables. This led to tests and confidence intervals for the difference of two means in Lesson 10.

In the case of standard deviations, it is the ratio of two random variables that yields the structure necessary for tests and confidence intervals. Hence we focus on the ratio of two populations standard deviations, rather than the difference. We get the same information concluding that the ratio is 1 as we do from concluding that a difference is 0. It does necessitate exploring still another probability distribution called the F family.

THE F-DISTRIBUTION

To perform a hypothesis test or obtain a confidence interval that is based on an F distribution, we will need to be able to find the F-value(s) corresponding to a specified area under an F-curve. Recall that the symbol F_α is used to represent the F-value with area α to its right under an F-curve. This quantity can be found with the aid of Minitab using the INVCDF command. In addition, we need to compute the area under an F curve in order to compute P-values for test statistics. We explain the details in the following examples.

EXAMPLE 11.5 Illustrates finding F percentiles

For an F-curve with df = (4, 12), find $F_{0.05}$. That is, find the F-value with area 0.05 to its right.

SOLUTION The command INVCDF with the subcommand F computes the F-value with a specified area to its *left*. We want to find the F-value having area 0.05 to its *right* for an F-curve with df = (4, 12). Since the total area under an F-curve is 1, this is equivalent to finding the F-value having area 0.95 to its *left* for an F-curve with df = (4, 12). Proceed as follows.

Menu commands:

1 Choose Calc ➤ Probability Distributions ➤ F...
2 Select the Inverse cumulative probability radio button
3 Select the Numerator degrees of freedom: text box and type 4
4 Select the Denominator degrees of freedom: text box and type 12
5 Select the Input constant: radio button
6 Press [Tab] and type 0.95 in the text box
7 Select [OK]

Session commands:

- Type `INVCDF 0.95;` at the `MTB >` prompt and press [Enter]
- Type `F 4 12.` at the `SUBC>` prompt and press [Enter]

Printout 11.10 on the following page displays the output.

PRINTOUT 11.10
Output of the F subcommand to INVCDF

```
F distribution with 4 DF in numerator and 12 DF in denominator

 P( X <= x)          x
     0.9500       3.2592
```

The output shows that for an F-curve with df = (4, 12), the F-value with area 0.95 to its left is 3.2592. Thus, for an F-curve with df = (4, 12), $F_{0.05} = 3.2592$. ■

As for finding the area under a given F-curve, the Minitab CDF command will accept the same subcommand as does INVCDF and this time returns *areas* to the left of a specified value.

EXAMPLE 11.6 Illustrates finding the area for specified F-values

For an F-curve with df = (4, 12), find the area to the right of $F = 1.8$.

SOLUTION The command CDF with the subcommand F computes the area for a specified area to the *left* of a given F-value. We want to find the area to the *right* of $F = 1.8$ with df = (4, 12). Proceed as follows.

Menu commands:

1 Choose Calc ➤ Probability Distributions ➤ F...
2 Select the Cumulative probability radio button
3 Select the Numerator degrees of freedom: text box and type 4
4 Select the Denominator degrees of freedom: text box and type 12
5 Select the Input constant: radio button
6 Press [Tab] and type 1.8 in the text box
7 Select [OK]

Session commands:

- Type CDF 1.8; at the MTB > prompt and press [Enter]
- Type F 4 12. at the SUBC> prompt and press [Enter]

Printout 11.11 displays the output.

PRINTOUT 11.11
Output of the F subcommand to CDF

```
F distribution with 4 DF in numerator and 12 DF in denominator

         x      P( X <= x)
    1.8000          0.8063
```

The output shows that for an F-curve with df = (4, 12), the area to the left of $F = 1.8$ is 0.8063 Thus, the area to the right of $F = 1.8$ is $1 - 0.8063 = 0.1937$ for an F-curve with df = (4, 12). ■

PROBLEM 11.5 For an F-curve with df = (12, 5), find
a) $F_{0.05}$ b) $F_{0.01}$ c) $F_{0.025}$
d) Find the area to the right of $F = 1.5$ under this curve ◇

THE F-TEST FOR TWO POPULATION STANDARD DEVIATIONS

Procedure 11.3 of *IS/6e* carries out the steps for and F-test of the hypothesis that two standard deviations are equal. Minitab has no set of commands for carrying out this procedure. However, some of the critical aspects of the test can be carried out with the assistance of commands in Minitab. We illustrate with an example from *IS/6e* .

EXAMPLE 11.7 Illustrates Procedure 11.3 for an F-test

In the 1997 paper "Using Repeatability and Reproducibility Studies to Evaluate a Destructive Test Method" (*Quality Engineering,* 10(2), pp.283–290), A. Phillips et al., studied the variability of the Elmendorf tear test. That test is used to evaluate material strength for fiberglass shingles, paper quality, and other manufactured products.

In one aspect of the study, the researchers independently and randomly obtained data on Elmendorf tear strength of three different vinyl floor coverings. Table 11.2 (WeissStats CD:Tb11-02) provides the data, in grams, for two of the three vinyl floor coverings.

TABLE 11.2 Results of Elmendorf tear test on two different vinyl floor coverings

Brand A		Brand B	
2288	2384	2592	2384
2368	2304	2512	2432
2528	2240	2576	2112
2144	2208	2176	2288
2160	2112	2304	2752

Apply Procedure 11.3 to test the hypothesis that the two population standard deviations are equal at the 5% level of significance using Minitab.

SOLUTION Enter the data into a Minitab worksheet in two columns named `A` and `B` in the usual manner. Now we apply the procedure in steps.

STEP 1 *State the null and alternative hypotheses.*

H_0: $\sigma_1 = \sigma_2$ (standard deviation of tear strengths are the same)
H_a: $\sigma_1 \neq \sigma_2$ (standard deviation of tear strengths are different)

Note that the test is two-tailed since there is a not-equal sign ($\neq$) in the alternative hypothesis.

STEP 2 *Decide on the significance level, α.*

The test is to be performed at the 5% significance level. Thus, $\alpha = 0.05$.

STEP 3 *The critical values for a two-tailed test are $F_{1-\alpha/2}$ and $F_{\alpha/2}$ with df $= (n_1 - 1, n_2 - 1)$.*

We have $\alpha = 0.05$ and $n_1 = 10$, $n_2 = 10$. Consequently, df $= (9, 9)$. Therefore, we need to determine $F_{1-0.025} = F_{0.975}$ and $F_{0.025}$for df $= (9, 9)$. The details are shown in Printout 11.12. You see from the printout that, for df =(9,9), we have $F_{0.975} = 0.24839$ and $F_{0.025} = 4.02599$

PRINTOUT 11.12
Finding F-values with Minitab

```
MTB > SET C10
DATA> .025 .975
DATA> END
MTB > INVCDF C10 C11;
SUBC> F 9 9.
MTB > PRINT C11
C11
 0.24893   4.02599
```

STEP 4 *Compute the value of the test statistic*

$$s_1^2/s_2^2$$

See Printout 11.13 for the execution.

PRINTOUT 11.13
Computing the test statistic with Minitab

```
MTB > LET K3=STDEV('A')**2/STDEV('B')**2
MTB > PRINT K3
K3    0.413028
```

Printout 11.13 shows that the value of the test statistic is

$$s_1^2/s_2^2 = 0.413028.$$

STEP 5 *If the value of the test statistic falls in the rejection region, reject H_0; otherwise, do not reject H_0.*

From Step 4, the value of the test statistic is $s_1^2/s_2^2 = 0.413$ to three decimal places. Since $0.24839 < 0.4130 < 4.02599$, we do not reject H_0. Alternatively, we can find the P-value of the test using Minitab as shown in Printout 11.14 on the following page.

PRINTOUT 11.14
Finding the *P*-value with Minitab

```
MTB > CDF 0.4130;
SUBC> F 9 9.

F distribution with 9 DF in numerator and 9 DF in denominator

          x       P( X <= x)
     0.4130           0.1019
```

From Printout 11.14, we see that the area to the left of the test statistic is 0.1019. Since the test is two-tailed, the *P*-value of the test is 2(0.1019) = 0.2038. Therefore, H_0 should not be rejected at the 5% level of significance.

STEP 6 *State the conclusion in words.*

The test results are not statistically significant at the 5% level, that is, the data do not provide sufficient evidence to conclude that the populations standard deviations of tear strength differ for the two vinyl floor coverings.

With all of the steps involved in Procedure 11.3 this is a natural place to construct a macro since Minitab does not provide a separate routine for this case. A macro entitled 2stdev.MAC has been included in the *WeissStats CD* for you. To run this macro, we assume it is located in the Macro folder of your Minitab installation. If not, adjust the calls in the following commands accordingly.

Menu commands:

1 Type %2stdev 'A' 'B'

2 Type Y in response to the question, Do you want to perform a hypothesis test (Y/N)? and press Enter↵

3 Type 0 in response to Enter 0, 1, or -1, respectively, for a two-tailed, right-tailed, or left-tailed test and press Enter↵

4 Type N in response to the question, Do you want a confidence interval (Y/N)? and press Enter↵

Printout 11.15 shows the resulting output, confirming the tedious calculations made above.

PRINTOUT 11.15
Output of the macro %2stdev

```
F-Test of sigma1 = sigma2 (vs not =)

Row  Variable    n     StDev       F      P

  1         A   10   128.322    0.41  0.204
  2         B   10   199.669
```

PROBLEM 11.6 Does cloud seeding makes a difference? In a study on the effects of cloud seeding in Tasmania, data were collected for various sections of the country at different seasons of the year. A partial list of reported rainfalls in inches for Eastern Tasmania during the summer season is summarized in the following table (S=seeded, U=unseeded).

S		U	
0.40	0.86	0.78	2.16
1.79	0.00	0.73	0.19
1.13	0.17	0.88	0.25
0.44	0.96	0.31	1.04
0.66	0.46	1.59	0.68
0.22	1.76	1.11	5.12

Is there statistical evidence to show that rainfall is more evenly spread out (less variable) in regions that were seeded over those that were not? Use the 5% level of significance and name your columns `S` and `U`.

CONFIDENCE INTERVALS FOR TWO POPULATION STANDARD DEVIATIONS

We can also use Minitab to aid us in constructing a confidence interval for two population standard deviations following the steps of Procedure 11.4 of *IS/6e* .

EXAMPLE 11.8 *Illustrates how Minitab can be used in conjunction with Procedure 11.4*

Refer to Example 11.7. Use the sample data in Table 11.2 to find a 95% confidence interval for the ratio, σ_1/σ_2, of the standard deviations of tear strength for Brand A and Brand B vinyl floor coverings. Employ Minitab in conjunction with Procedure 11.4 of *IS/6e* .

SOLUTION We repeat the steps of Procedure 11.4 here.

STEP 1 *With confidence level of* $1-\alpha$, *find* $F_{1-\alpha/2}$ *and* $F_{\alpha/2}$ *for the F-curve with* $df = (n_1 - 1,\ n_2 - 1)$.

From Step 3 of Example 11.3, $F_{0.975} = 0.24839$ and $F_{0.025} = 4.02599$.

STEP 2 *The confidence interval for* σ_1/σ_2 *is*

$$\frac{1}{\sqrt{F_{\alpha}/2}} \cdot \frac{s_1}{s_2} \quad \textit{to} \quad \frac{1}{\sqrt{F_{1-\alpha/2}}} \cdot \frac{s_1}{s_2}$$

From Printout 11.15 we know that $s_1 = 128.322$ and $s_2 = 199.669$. Thus, the confidence interval is from

$$\frac{1}{\sqrt{4.02599}} \cdot \frac{128.322}{199.669} \quad \textit{to} \quad \frac{1}{\sqrt{0.24839}} \cdot \frac{128.322}{199.669}$$

or 0.32 to 1.28. We can be 95% confident that the ratio of standard deviations of tear strength for Brand A and Brand B vinyl floor coverings is somewhere between 0.32 and 1.28.

You will recall that the %2stdev macro prompted for a confidence interval at Step 5. You may use the macro to compute only a confidence interval as follows.

Menu commands:

1 Type `%2stdev 'A' 'B'`
2 Type N in response to the question, Do you want to perform a hypothesis test (Y/N)? and press [Enter]
3 Type Y in response to the question, Do you want a confidence interval (Y/N)? and press [Enter]
4 Type 95 in response to Enter the confidence level, as a percentage and press [Enter]

Printout 11.16 shows the resulting output, confirming the calculations made above.

PRINTOUT 11.16
Confidence interval using %2stdev

```
Row  Variable    n    StDev  Level     CI for ratio

  1         A   10  128.322  95.0%  (0.320, 1.290)
  2         B   10  199.669
```

To test the null hypothesis using the 95% confidence interval, we fail to reject the null hypothesis since the hypothesized ratio of 1 falls inside the confidence interval, that is, $.32 < 1 < 1.290$. ■

PROBLEM 11.7

Refer to Problem 11.6. Determine a 90% confidence interval for the ratio of standard deviations. Apply the confidence interval method for the hypothesis test in Problem 11.6. ◇

LESSON 12

Inferences for Population Proportions

GENERAL OBJECTIVE

In Lessons 8–10 we examined confidence intervals and hypothesis tests for population means. Now we will learn how to conduct those inferences for population proportions, which are themselves population means.

LESSON OUTLINE

12.1 Confidence intervals for one population proportion

E-11.1:I-12.1

Procedure 12.1 in *IS/6e* (Procedure 11.1 in *ES/5e*) provides a method for obtaining a large-sample confidence interval for a population proportion, called the One-sample z-interval procedure. It assumes that the sample size is large, defined as having both the number of successes and the number of failures are both greater than 5 (10 might be better). Minitab implements this formula under Basic Statistics in the general Stat menu, as well as by the session command PONE. An example of implementing this command follows.

EXAMPLE 12.1 *Illustrates the z-interval procedure in Minitab*

A poll was taken of 1010 U.S. employees. The employees sampled were asked whether they "play hooky," that is, call in sick at least once a year when they simply need time to relax; 202 responded in the affirmative. Use these data and the PONE command to obtain a 95% confidence interval for the proportion, p, of all U.S. employees who play hooky.

SOLUTION The sample size of 1010 is certainly large enough to call for the z-interval. Typically, the data are in summary form as in this example and Minitab allows for this as well as for having the original data (the yes's and no's or 1's and 0's) in a column. Proceed as follows.

Menu commands:

1. Choose Stat ➤ Basic Statistics ➤ 1 Proportion...
2. Select the Summarized data radio button
3. Select the Number of trials: text box and type 1010
4. Select the Number of successes: text box and type 202
5. Click the [Options...] button and type 95 in the Confidence level: text box
6. Select the Use test and interval based on normal distribution check box
7. Select [OK] twice

Printout 12.1 displays the result.

PRINTOUT 12.1 Output of the z-interval procedure in Minitab

```
Test of p = 0.5 vs p not = 0.5

Sample     X      N  Sample p          95.0 % CI          Z-Value  P-Value
1        202   1010  0.200000  (0.175331, 0.224669)    -19.07    0.000
```

The same result can be achieved with the following session commands even in earlier releases.

Session commands:

- Type PONE 1010 202; at the MTB > prompt and press [Enter]
- Type Confidence 0.95 at the SUBC> prompt and press [Enter]
- Type UseZ. at the SUBC> prompt and press [Enter]

In the columns headed X, N, and Sample p, you will find the number of successes, the sample size, and the calculated sample proportion, $\hat{p}$. Then you find that the 95% confidence interval for p runs from 0.1753331 to 0.2246669. We can be 95% confident that the proportion of all employees who play hooky is somewhere between 17.5% and 22.5%. Notice that a hypothesis test appears to be carried out as well. You may ignore that part of the output for the present; we will take up that matter in the next section.

Of course, the z-interval computed this way is only approximate since the distribution of $\hat{p}$, a discrete random variable, cannot be exactly normal no matter what the value of n. For small sample sizes, it would not be advisable to use the z-procedure. Minitab supports a more exact procedure based on the discrete distribution, called the exact z-interval. The formulas are quite complicated but the results may be viewed by making certain that the Use test and interval based on normal distribution is *NOT* checked, that is, cleared. In the session commands, simply omit the subcommand UseZ.

EXAMPLE 12.2 *Illustrates the exact interval procedure in Minitab*

Using the data of Example 12.1, find the exact 95% confidence interval for the proportion, p, of all U.S. employees who play hooky.

SOLUTION For this example, the output appears in Printout 12.2.

PRINTOUT 12.2
Output of the exact procedure in Minitab

```
Test of p = 0.5 vs p not = 0.5

                                                       Exact
Sample       X       N  Sample p          95.0 % CI       P-Value
1          202    1010  0.200000  (0.175739, 0.226023)     0.000
```

Not surprisingly, the results are quite similar to those previously obtained because of the large sample size. Nevertheless, there are differences and this procedure should be used whenever the sample is small. ■

PROBLEM 12.1

A 1985 Gallup Poll estimated the support among Americans for "right to die" laws. For the survey, 1528 adults, 18 years old and older, were asked whether they were in favor of voluntary withholding of life support systems from the terminally ill. The results: 1238 said they were. Find a 99% confidence interval for the proportion of all adult Americans who are in favor of "right to die" laws. ◇

PROBLEM 12.2 According to genetic theory, one-fourth of the offspring of certain kinds of mice should be white, the remainder should be black or spotted. In three litters, two of these mice have produced 21 mice, 9 of which are white. Does this show the genetic model is in error? ◇

12.2 Hypothesis tests for one population proportion

E-11.2:I-12.2

As you have already observed, Minitab carries out a hypothesis test as part of its PONE command. You have only to input the hypothesized proportion. To perform a large-sample hypothesis test with null hypothesis H_0: $p = p_0$, we can use the test statistic

$$z = \frac{\hat{p} - p_0}{\sqrt{p_0(1 - p_0)/n}}$$

a z-test. Let us illustrate this application using another example from *IS/6e* and *ES/5e*.

EXAMPLE 12.3 *Illustrates the use of PONE*

One of the more controversial issues in the United States is that of gun control; there are many avid proponents and opponents of banning handgun sales. In an April 1993 survey, Louis Harris of LH Research Inc. polled 1250 U.S. adults regarding their view on banning handgun sales. The results were that 650 of those sampled favored a ban. Do the data provide sufficient evidence to conclude that a majority of U.S. adults (i.e., more than 50%) favor banning handgun sales?

SOLUTION According to the analysis of the problem in *IS/6e* and *ES/5e*, we should perform a right tailed test of $H_0 : p = 0.5$ at the 5% level of significance. Proceed as follows.

Menu commands:

1 Choose Stat ➤ Basic Statistics ➤ 1 Proportion...
2 Select the Summarized data option button
3 Select the Number of trials: text box and type 1250
4 Select the Number of successes: text box and type 650
5 Click the [Options...] button and type 90 in the Confidence level: text box
6 Select the Test proportion: text box and type 0.5
7 Press the [▾] button in the Alternative: text box and select greater than from the drop down list
8 Select the Use test and interval based on normal distribution check box
9 Select [OK] twice

Session commands:

- Type PONE 1250 650; at the MTB > prompt and press Enter
- Type Test 0.5; at the SUBC> prompt and press Enter
- Type Alternative 1; at the SUBC> prompt and press Enter
- Type UseZ. at the SUBC> prompt and press Enter

Printout 12.3 displays the results of these commands.

PRINTOUT 12.3
Output of the z-interval procedure in Minitab

```
Test of p = 0.5 vs p > 0.5

Sample      X     N  Sample p  95.0% Lower Bound  Z-Value  P-Value
1         650  1250  0.520000           0.496757     1.41    0.079
```

With a P-value of $0.079 > 0.05$, there is not enough evidence to reject the null hypothesis. A lower one-sided 95% confidence interval for p is $(0.497, 1)$ which is further evidence for not rejecting the null hypothesis since this interval includes the value 0.5. ■

PROBLEM 12.3 According to the Arizona Real Estate Commission, in 1982, only 10% of the people holding real estate licenses were active in the industry. An independent agency has been asked by the Commission to determine whether this year's percentage is higher. A random sample of 150 licensed people reveals that 24 are currently active. Perform the appropriate hypothesis test at the 1% significance level. ◇

12.3 Inferences for two population proportions using independent samples

E-11.3:I-12.3

Now we will examine methods for performing inferences to compare the proportions, p_1 and p_2, of two populations when we have large sample sizes for both samples and the samples are independent. Minitab has a command, PTWO, able to achieve both confidence intervals and tests at the same time.
The test of the null hypothesis

$$H_0\colon p_1 = p_2 \text{ (population proportions are equal)}$$

may be carried out as a z-test using the test statistic

$$z = \frac{\hat{p}_1 - \hat{p}_2}{\sqrt{\hat{p}_p(1-\hat{p}_p)}\sqrt{(1/n_1)+(1/n_2)}}$$

where

$$\hat{p}_p = \frac{x_1 + x_2}{n_1 + n_2}$$

is the pooled sample proportion, an estimate of $p = p_1 = p_2$ when H_0 is true. This is summarized as Procedure 12.3 in *IS/6e* (Procedure 11.3 in *ES/5e*) and called the *Two-sample z-test* .

The endpoints of the confidence interval corresponding to the choice α are

$$(\hat{p}_1 - \hat{p}_2) \pm z_{\alpha/2} \cdot \sqrt{\hat{p}_1(1-\hat{p}_1)/n_1 + \hat{p}_2(1-\hat{p}_2)/n_2}$$

These formulas are summarized as Procedure 12.4 in *IS/6e* (Procedure 11.4 in *ES/5e*) and is called the *Two-sample z-interval.* Let us implement the Minitab command PTWO to solve problems involving two proportions with large sample sizes.We illustrate using Example 12.8 in *IS/6e* (Example 11.8 in *ES/5e*).

EXAMPLE 12.4 *Illustrates the use of PTWO in Minitab*

A *Zogby International* poll of 1181 adults nationwide was conducted in March, 1999 to gauge the demand for vegetarian meals in restaurants. The study was commissioned by the Vegetarian Resource Group and was published in the September/October 1999 issue of the *Vegetarian Journal.*

In the survey, independent random samples of 747 U.S. men and 434 U.S. women were taken. Of those sampled, 276 men and 195 women said that they sometimes ordered a dish without mean, fish or fowl when they eat out. Do the data provide sufficient evidence to conclude that, in the United States as a whole, the percentage of men is smaller than the percentage of women who so order? Use a 5% level of significance.

SOLUTION Let p_1 denote the proportion of all U.S. men who sometimes order vegetarian and let p_2 denote the proportion of all U.S. women who so order. Then the null and alternative hypotheses are

H_0: $p_1 = p_2$ (percentage for men is not less than that for women)

H_a: $p_1 < p_2$ (percentage for men is less than that for women)

Note that the hypothesis test is left-tailed since there is a less-than sign (<) in the alternative hypothesis. Proceed as follows.

Menu commands:

1 Choose Stat ➤ Basic Statistics 2 Proportions...

2 Click the Summarized data: option button

3 Press Tab and type 747 in the First sample: text box under Trials:

4 Press Tab and type 276 in the First sample: text box under Successes:

5 Press Tab and type 434 in the Second sample: text box under Trials:

6 Press Tab and type 195 in the Second sample: text box under Successes:

7 Click the [Options...] button

8 Type 95 in the Confidence level: text box

9 Press [Tab] and type 0 in the Test difference: text box

10 Press the [▼] button in Alternative: text box and select less than from the drop-down list

11 Select the Use pooled estimate of p for test check box

12 Select [OK] twice

Session commands:

- Type `PTWO 747 276 434 195;` at the `MTB >` prompt and press [Enter]
- Type `Test 0;` at the `SUBC>` prompt and press [Enter]
- Type `Pooled.` at the `SUBC>` prompt and press [Enter]

Printout 12.4 shows the results of applying these commands to this problem.†

PRINTOUT 12.4
Output of PTWO

```
Sample       X      N  Sample p
1          276    747  0.369478
2          195    434  0.449309

Estimate for p(1) - p(2):  -0.0798308
95% upper bound for p(1) - p(2):  -0.0309817
Test for p(1) - p(2) = 0 (vs < 0):  Z = -2.70  P-Value = 0.003
```

From the summary data, single estimates are given for the two proportions as well as an estimate for the difference. With a P-value of 0.003, you would certainly reject the null hypothesis at the 5% level of significance. Since this is a left-tailed test, the single critical value is -1.645 (type `INVCDF 0.05` at the `MTB >` prompt). Since the value of the test statistic, -2.70 is less than this value, you would reject the null hypothesis using a classical test. Finally, because the test is left-tailed we need an upper one-sided confidence interval for the difference, p. From the printout you see that the 95% upper bound is -0.0309817 so that the upper one-sided 95% confidence interval is $(-\infty, -0.031)$. Since 0 is not in this interval, you would reject as before.

At the 5% level of significance, the data provide sufficient evidence to conclude that, in the United States as a whole, the proportion of men who sometimes order vegetarian is smaller that the proportion of women who sometimes so order. ■

You may of course use PTWO for computing two-sided confidence intervals for any set of data. Just be sure to input the desired confidence level at Step 8 and select not equal as an alternative choice. Then ignore the output having to do with hypothesis tests.

† Use the pooled option only when the test difference is 0.

PROBLEM 12.4 The Organization for Economic Cooperation and Development, Paris, France, summarizes data on labor-force participation rates in *Labour Force Statistics*. From independent samples of 300 U.S. women and 250 Canadian women it is found that 184 of the U.S. women and 148 of the Canadian women are in their respective labor forces. At the 5% significance level, do the data suggest that there is a difference between the labor-force participation rates of U.S. and Canadian women? Test this using all three methods. ◇

LESSON 13

Chi-square Procedures

GENERAL OBJECTIVE

In previous lessons, we have examined the Minitab commands that are used for statistical inferences concerning population means. Now we will learn how Minitab can be applied to perform *chi-square procedures.* Specifically, we will study the Minitab commands required to carry out a chi-square goodness-of-fit test, and a chi-square independence test.

LESSON OUTLINE

13.1 Chi-square goodness-of-fit test

E-12.2:I-13.2

As pointed out in your textbook,the statistical inference techniques presented so far have dealt exclusively with hypothesis tests and confidence intervals for specific population parameters, such as population means, populations standard deviations and population proportions. However, there are many situations where we are concerned, not with specific parameters but general situations regarding the status quo versus a change.

The inferential procedures discussed in this lesson rely on a class of continuous probability distributions called *chi-square distributions.* We pause to study that family of distributions.

THE χ^2-DISTRIBUTION

Probabilities for a random variable having a chi-square distribution are equal to areas under a curve–a χ^2 `(chi-square) curve`(see Figure 13.1).

FIGURE 13.1
A typical χ^2-curve

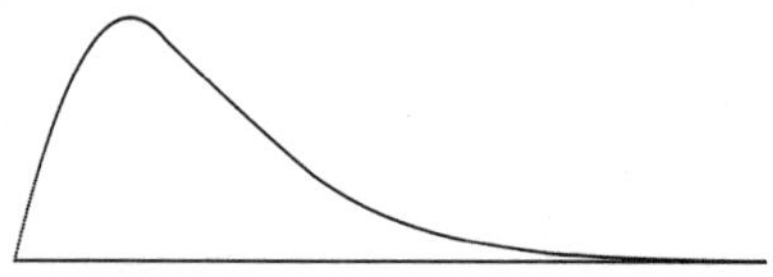

To perform a hypothesis test or obtain a confidence interval that is based on a chi-square distribution, we will need to be able to find the χ^2-value(s) corresponding to a specified area under a χ^2-curve. Recall that the symbol χ^2_α is used to represent the χ^2-value with area α to its right under a χ^2-curve. This quantity can be found using Table V in *IS/6e* and *ES/5e* or, alternatively, with the aid of Minitab. In addition, we need to compute the area under a χ^2 (chi-square) curve in order to compute P-values for test statistics.

Up to now the distributions we have been dealing with for statistical inference have been symmetric. That is, the area to the right of the mean is a mirror image of the area to the left. This is not true with the χ^2 family of distributions (see Figure 13.1). Like the t family, each χ^2 distribution is distinguished by a parameter known as degrees of freedom, a positive whole number that we will have to supply to Minitab in order to produce a desired output.

Minitab has a subcommand called CHISQUARE which, with a specification of the degrees of freedom, may be used with both the CDF and the INVCDF commands. In conjunction with the INVCDF command, this subcommand may be used to compute the χ^2-value with a specified area to its *left.* From this, it is a simple matter to obtain the χ^2-value(s) corresponding to any specified area under an arbitrary χ^2-curve. We present the details in the example to follow.

EXAMPLE 13.1 *Illustrates how to find the χ^2-value for a specified area using Minitab*

For a χ^2-curve with 20 degrees of freedom, find $\chi^2_{0.05}$; that is, find the χ^2-value with area 0.05 to its right. See Figure 13.2.

FIGURE 13.2
The χ^2-curve with 20 degrees of freedom

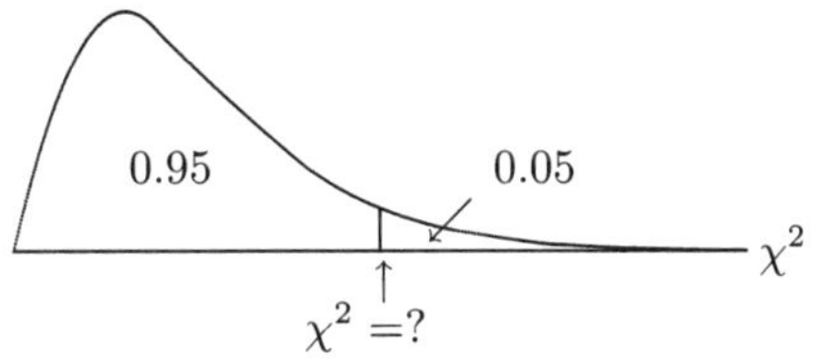

SOLUTION To find the χ^2-value with area 0.05 to its right, we observe that the area to the left must be 0.95. Proceed as follows.

Menu commands:

1 Choose Calc ➤ Probability Distributions ➤ Chi-square...
2 Select the Inverse cumulative probability radio button
3 Select the Degrees of freedom: text box and type 20
4 Select the Input constant: radio button
5 Press [Tab] and type 0.95 in the text box
6 Select [OK]

Session commands:

- Type `INVCDF 0.95;` at the `MTB >` prompt and press [Enter]
- Type `CHISQUARE 20.` at the `SUBC>` prompt and press [Enter]

PRINTOUT 13.1
Finding a χ^2-value

```
Chi-Square with 20 DF

 P( X <= x)          x
    0.9500     31.4104
```

From the resulting printout, we see that $\chi^2_{0.05} = 31.4104$. ■

PROBLEM 13.1

For a χ^2-curve with 14 degrees of freedom, use Minitab to find
a) the χ^2-value with area 0.10 to its left.
b) the χ^2-value with area 0.10 to it right. ◇

To calculate P-values, we need to be able to find the area under a χ^2-curve for given values. Using the CDF command in Minitab, we can find such areas.

EXAMPLE 13.2 *Illustrates finding the area under a χ^2-curve using Minitab*

For a χ^2-curve with 20 degrees of freedom, find the area to the right of the value $\chi^2 = 23$. See Figure 13.3.

FIGURE 13.3
The χ^2-curve with 20 degrees of freedom

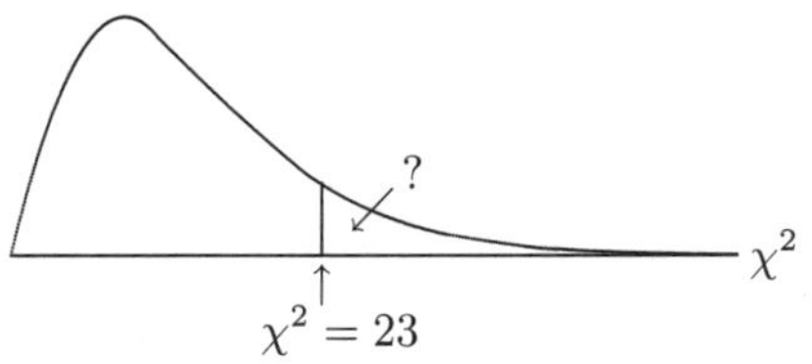

SOLUTION Recall that the CDF command in Minitab always produces the area to the *left* of any value that we specify. Hence we will need to find that area for the χ^2 value of 23 first. Proceed as follows.

Menu commands:

1 Choose Calc ➤ Probability Distributions ➤ Chi-square...
2 Select the Cumulative probability radio button
3 Select the Degrees of freedom: text box and type 20
4 Select the Input constant: radio button
5 Press Tab and type 23 in the text box
6 Select OK

Session commands:

- Type CDF 23; at the MTB > prompt and press Enter
- Type CHISQUARE 20. at the SUBC> prompt and press Enter

The result of either of these commands is displayed in Printout 13.2.

PRINTOUT 13.2
Finding a χ^2-value

```
Chi-Square with 20 DF

          x      P( X <= x)
    23.0000          0.7112
```

From the printout, we see that $P(\chi^2 \leq 23) = 0.7112$. Hence the area to the right of $\chi^2 = 23$ is approximately 0.29. ■

PROBLEM 13.2 For a χ^2-curve with 14 degrees of freedom, use Minitab to find
a) the area to the right of $\chi^2 = 13$.
b) the area between $\chi^2 = 7$ and $\chi^2 = 13$. ◇

13.2 Chi-square goodness-of-fit test

E-12.2:I-13.2

The first chi-square procedure we will examine is called the *chi-square goodness-of-fit test.* Among other things, this procedure can be used to test hypotheses about the percentage distribution of a population or the probability distribution of a random variable.

Minitab does not have a special command to perform a chi-square goodness-of-fit test. Nonetheless, we can still use Minitab to help with the computations required to carry out such a test. In the next example we will apply Procedure 13.1 starting of *IS/6e* (Procedure 12.1 of *ES/5e*) to perform a chi-square goodness-of-fit test. The calculations required will be done by Minitab.

EXAMPLE 13.3 *Illustrates how Minitab can be used for a goodness-of-fit test*

The U.S. Bureau of Investigation (FBI) compiles data on crimes and crime rates and publishes the information in *Crime in the United States.* A violent crime is classified by the FBI as murder, forcible rape, robbery, or aggravated assault. Table 13.1 provides a relative-frequency distribution for (reported) violent crimes in 1995.

TABLE 13.1 Distribution of violent crimes in the United States, 1995

Type	Relative frequency
Murder	0.012
Forcible rape	0.054
Robbery	0.323
Agg. assault	0.611
	1.000

A random sample of 500 violent-crime reports from last year yielded the frequency distribution shown in Table 13.2.

TABLE 13.2 Sample results for 500 randomly selected violent-crime reports from last year

Type	Frequency
Murder	9
Forcible rape	26
Robbery	144
Agg. assault	321
	500

At the 5% significance level, do the data provide sufficient evidence to conclude that last year's distribution of violent crimes has changed from the 1995 distribution?

SOLUTION In applying Procedure 13.1 of *IS/6e* (Procedure 12.1 of *ES/5e*) to this example, we will repeat all of the steps using Minitab to carry out the computations and display the results.

STEP 1 *State the null and alternative hypotheses.*

The null and alternative hypotheses are

H_0:Last year's violent-crime distribution is the same as the 1995 distribution

H_a:Last year's violent-crime distribution is different from the 1995 distribution

STEP 2 *Calculate the expected frequencies using the formula*

$$E = np$$

where n denotes the sample size and p denotes the probability for the category given in the null hypothesis.

We can use Minitab to determine the expected frequencies by proceeding in the following manner. First enter the probabilities for the four specialties (categories) into a column named p. We will store the expected frequencies in a column named E. The expected frequencies can now be obtained by multiplying the probabilities in p by the sample size of $n = 500$. We have Minitab do that and store the expected frequencies in E by typing `LET 'E'=500*'p'`. Then type `PRINT 'E'` to have Minitab print the expected frequencies. See Printout 13.3 for the results.

PRINTOUT 13.3
Finding expected frequencies with Minitab

```
E
    6.0    27.0    161.5    305.5
```

STEP 3 *Check whether the expected frequencies satisfy Assumptions 1 and 2, namely, all expected frequencies are at least 1 and at most 20% of the expected frequencies are less than 5.*

1. All expected frequencies are at least 1? Yes (see Printout 13.3).
2. At most 20% of the expected frequencies are less than 5? Yes, because there are no expected frequencies less than 5.

STEP 4 *Decide on the significance level, α.*

We are to perform the test at the 5% significance level. Thus, $\alpha = 0.05$.

STEP 5 *The critical value is χ^2_α with df $= k - 1$, where k is the number of categories in the distribution.*

From Step 4, $\alpha = 0.05$. Also, there are four categories (specialties), so $k = 4$. Thus, we need to find $\chi^2_{0.05}$ for a χ^2-curve with

$$\text{df} = k - 1 = 4 - 1 = 3$$

We learned how to do this using Minitab in the previous section. See Printout 13.4 on the following page for the output.

PRINTOUT 13.4
Finding a χ^2-value

```
Chisquare with 3 d.f.

 P( X <= x)           x
    0.9500       7.8147
```

The printout shows that, for df = 3, $\chi^2_{0.05} = 7.8147$. This is the required critical value. The

STEP 6 *Compute the value of the test statistic*

$$\chi^2 = \Sigma(O - E)^2/E$$

Here we can use primitive session commands to construct the value of this test statistic. We begin by storing the observed frequencies, given in the first column of Table 13.2 into a column named O. Recalling that the expected frequencies are stored in E, we can now obtain the value of the test statistic, χ^2, by typing `LET K1=SUM(('O'- 'E')**2/'E')` followed by `PRINT K1` at the `MTB >` prompt. See Printout 13.5 for the results.

PRINTOUT 13.5
Computing the test statistic with Minitab

```
 MTB > LET K1=SUM(('O'-'E')**2/'E')
 MTB > PRINT K1
 K1    4.21974
```

The last item in Printout 13.5 shows that the value of the test statistic is

$$\chi^2 = \Sigma(O - E)^2/E = 4.21974$$

STEP 7 The procedure calls for rejecting H_0 only if the value of the test statistic is too large; that is, the test is always a right-tailed test with rejection region (χ^2_α, ∞); otherwise, do not reject H_0.

From Step 6, we see that the value of the test statistic is $\chi^2 = 4.21974$. Comparing this with the critical value of $\chi^2_{0.05} = 7.8147$ found in Step 5, we see that the value of the test statistic does *not* fall in the rejection region. Hence, we do not reject H_0.

Alternatively, since the test is right-tailed, we may find the P-value by subtracting the value obtained by applying the CDF command at 4.21974. Printout 13.6 shows the necessary steps.

PRINTOUT 13.6
The P-value of the test

```
 MTB > CDF 4.82575;
 SUBC> CHISQUARE 3.
 Chi-Square with 3 DF

         x    P( X <= x )
    4.2197         0.7613
```

From the printout we see the P-value for this test is $1 - 0.7613 = 0.2387$. We reach the same conclusion by observing that the P-value of 0.2387 is greater than 0.05. Indeed, we would not reject at the 20% level of significance either.

STEP 8 *State the conclusion in words.*

The sample data do not provide sufficient evidence to conclude that last year's violent-crime distribution has changed from the 1995 distribution. ■

You may also use the % macro called fittest.mac provided on your accompanying *WeissStats CD*. Copy that file to the Macros folder of your Minitab installation and type %fittest at the MTB > prompt and press Enter. Following the instructions for input to that macro, the output is shown in Printout 13.7.

PRINTOUT 13.7
The P-value of the test

```
Chi-square goodness-of-fit test

Row      n     k     ChiSq     P-value

  1    500     4   4.21974    0.238693
```

Of course, the results agree. ■

PROBLEM 13.3

According to the document *Current Housing Reports,* published by the U.S. Bureau of the Census, the primary-heating-fuel distribution for occupied housing units is as follows:

Primary heating fuel	Percent
Natural gas	56.7
Fuel oil, kerosene	14.3
Electricity	16.0
LPG	4.5
Wood	6.7
Other	1.8

A random sample of 250 occupied housing units built after 1974 yields the frequency distribution below.

Primary heating fuel	Frequency
Natural gas	91
Fuel oil, kerosene	16
Electricity	110
LPG	14
Wood	17
Other	2

Do the data provide sufficient evidence to conclude that the primary-heating-fuel distribution for occupied housing units built after 1974 differs from that of all occupied housing units? Use $\alpha = 0.05$. ◇

13.3 Contingency tables; association

E-12.3:I-13.3

Sometimes surveys are initiated in order to determine whether or not there is a relationship between two variables that are classified in two ways. Indeed, many times the observed values of these variables are not even numerical. Yet we can determine the frequency of occurrence in each classification and produce a two-way table called a *contingency table*. Such tables now have numerical entries and can then be analyzed by statistical methods.

CONTINGENCY TABLES

The raw data for a contingency analysis may be in the form of simply a tabulation of responses to a survey or a classification into two or more nonnumeric categories. You may find such data in many sources including the Internet. Such data can be input directly into a Minitab worksheet to produce a contingency table and proceed to the analysis. Of course such data usually contain hundreds if not thousands of observations and hence we look to the computer for converting such raw data into a contingency table.

EXAMPLE 13.4 *Responses by sex to a questionnaire*

Here is an example of an abbreviated (for the sake of convenience) table of responses to a survey asking respondents whether or not they approve of the way Congress currently carries out its business. Respondents were only allowed to answer with Yes, No, or Und (for undecided). The results appear in the Table 13.3 on the following page, arranged in matching pairs, (Sex, Reply). You must keep this pairing in mind whenever using the data that are registered here.† Convert these data into a two-way contingency table using Minitab.

SOLUTION First enter the above data as text data in two columns named SEX and REPLY into a Minitab workspace in the usual manner. Then proceed as follows.

Menu commands:

1 Choose Stat ➤ Tables ➤ Cross Tabulation...

2 Enter SEX and REPLY in the Classification variables: text box

3 Select the Counts check box in the Display list and be sure the other choices are deselected (not checked or cleared)

† If the frequency of occurrence of each category is already summarized, see the next section for the method of data entry.

TABLE 13.3 Survey on Congressional approval

Sex	Reply	Sex	Reply
Male	Yes	Female	No
Male	No	Male	No
Female	Yes	Male	No
Male	Und	Female	Yes
Male	No	Male	Und
Female	Yes	Female	No
Female	Und	Female	Yes
Male	Yes	Male	No
Male	No	Male	Und
Male	Yes	Male	Und
Female	No	Male	No
Male	Yes	Male	No
Male	No	Male	No
Female	No	Male	Und
Male	No	Male	Und
Male	Yes	Female	No
Female	No	Female	Und
Female	Yes	Female	Yes
Female	Yes	Female	No
Female	Und	Female	Und

4 Clear (deselect) the Frequencies are in: check box and click [OK]

Session commands:

- Type Table 'SEX' 'REPLY'; at the MTB > prompt and press [Enter]
- Type Counts. at the SUBC> prompt and press [Enter]

The output of either of these two sets of commands is displayed in Printout 13.8.

PRINTOUT 13.8 Minitab contingency table for Sex and Reply

```
Rows: SEX      Columns: REPLY

               No       Und       Yes       All

Female          7         4         7        18
Male           11         6         5        22
All            18        10        12        40

 Cell Contents --
                Count
```

Totals are labeled All by Minitab. Each cell contains the frequency of respondents in the joint classification of the row and column marking that cell. Row totals give summaries of the number of respondents in each Sex category, while column totals summarize by the Reply category. ■

PROBLEM 13.4

Convert the contingency table output of Example 13.4 to percentages (hence probabilities) by selecting Total percents instead of Counts at Step 3 of the Menu commands. (TotPercents as the subcommand in the Session commands.) What are the chances that a respondent is both Male and answers Yes? ◇

ASSOCIATION

To examine the association between Sex and Reply in the population, we need to construct and examine a conditional probability distribution as defined and discussed in *IS/6e* and *ES/5e*.

EXAMPLE 13.5 Association between Sex and Reply

Refer to Example 13.4. Use Minitab to examine the relationship between the two variables Sex and Reply.

SOLUTION We will use Minitab to construct a conditional distribution of Sex by Reply using the data from Example 13.4. Proceed as follows.

Menu commands:

1. Choose Stat ➤ Tables ➤ Cross Tabulation...
2. Enter SEX and REPLY in the Classification variables: text box
3. Select the Column percents check box in the Display list and be sure the other choices are deselected (not checked)
4. Select [OK]

Session commands:

- Type Table 'SEX' 'REPLY'; at the MTB > prompt and press [Enter↵]
- Type Colpercents. at the SUBC> prompt and press [Enter↵]

The output is shown in Printout 13.9.

PRINTOUT 13.9
Minitab conditional distribution of Sex by Reply

```
Rows: SEX      Columns: REPLY

              No       Und       Yes       All

Female     38.89     40.00     58.33     45.00
Male       61.11     60.00     41.67     55.00
All       100.00    100.00    100.00    100.00
  Cell Contents --
                      % of Col
```

For example, there is a 45% chance that a respondent is female but if we knew the response to the questionnaire was Und, there is only a 40% chance that it came from a female. There appears to be an association between these two variables. ■

PROBLEM 13.5 Using the outputs of this section, practice drawing your own conclusions for various choices of variables. ◇

13.4 Chi-square independence test

E-12.4:I-13.4

In the last section, we restricted our conclusions to the hypothetical condition that the data represented a population. However, in almost all cases, these surveys represent a sample of the entire population, since data for the entire population are usually not available. Thus, in most cases, we must employ inferential statistics to decide whether two characteristics of a population are statistically dependent.

One of the most commonly used tests for statistical dependence is the *chi-square independence test.* The steps for carrying out this test are given as Procedure 13.2 in *IS/6e* (Procedure 12.2 in *ES/5e*). The computations required are generally tedious and time-consuming. Fortunately, Minitab has a command, called CHISQUARE, that will perform Procedure 13.2 for us. The next example shows how CHISQUARE is applied when data are available in contingency table.

EXAMPLE 13.6 *Illustrates the use of CHISQUARE*

A national survey was conducted to obtain information on the alcohol consumption patterns of American adults by marital status. A random sample of 1772 residents, 18 years old and over, yielded the tabulated data shown in Table 13.4† (WeissStats CD:Tb13-13).

TABLE 13.4
Contingency table of marital status versus alcohol consumption for 1772 randomly selected adults

Status	Drinks per month: Abstain	1–60	Over 60	Total
Single	67	213	74	354
Married	411	633	129	1173
Widowed	85	51	7	143
Divorced	27	60	15	102
Total	590	957	225	1772

Do the data suggest, at the 5% significance level, that marital status and alcohol consumption patterns are statistically dependent? As pointed out in *IS/6e* and *ES/5e* the null hypothesis is that marital status and alcohol consumption are statistically *independent*. Rejection of that hypothesis leads to the conclusion that

† Adapted from: Clark, W. B. and Midanik, L. "Alcohol Use and Alcohol Problems among U.S. Adults: Results of the 1979 National Survey." In: National Institute on Alcohol Abuse and Alcoholism. Alcohol and Health Monograph No 1, Alcohol Consumption and Related Problems. DHHS Pub. No (ADM) 82–1190, 1982.

there is a relationship between marital status and alcohol consumption. Since the data represent only a sample of individuals and not the entire population, we cannot examine the question with conditional distributions as we did in the last section.

SOLUTION We want to perform the hypothesis test

H_0: Marital status and alcohol consumption are statistically independent.

H_a: Marital status and alcohol consumption are statistically dependent.

at the 5% significance level.

To employ Minitab, first enter the data into the computer in columns named Abstain, 1-60, Over 60. That is, Abstain should contain the column data for the number in each of the four marital status categories who abstain, 1-60 should contain the number who drink 1–60 drinks per month, Over 60 should contain the number who drink over 60 drinks per month. Then, proceed as follows.

Menu commands:

1 Choose Stat ➤ Tables ➤ Chisquare Test...

2 Enter Abstain 1-60 Over 60 in the text box labeled Columns containing the table:

3 Select [OK]

Printout 13.10 displays the results.

PRINTOUT 13.10 Output of CHISQUARE

```
Expected counts are printed below observed counts

       Abstain     1-60  Over 60    Total
    1       67      213       74      354
        117.87   191.18    44.95

    2      411      633      129     1173
        390.56   633.50   148.94

    3       85       51        7      143
         47.61    77.23    18.16

    4       27       60       15      102
         33.96    55.09    12.95

Total      590      957      225     1772

Chi-Sq = 21.952 +  2.489 + 18.776 +
          1.070 +  0.000 +  2.670 +
         29.358 +  8.908 +  6.856 +
          1.427 +  0.438 +  0.324 = 94.269
DF = 6, P-Value = 0.000
```

The same results may be achieved with the following session commands.

Session commands:

- Type `CHISQUARE 'Abstain' '1-60' 'Over60'` at the `MTB >` prompt and press Enter

The output in Printout 13.10 first displays a table of observed and expected frequencies. This is Minitab's version of Table 13.14 in *IS/6e* (and Table 12.14 in *ES/5e*). The expected frequency for each cell is placed beneath the corresponding observed frequency. Thus, for instance, the table shows that the observed frequency of the adults sampled who are single and have over 60 drinks per month is 74, whereas if marital status and alcohol consumption are statistically independent, then the expected frequency is computed to be 44.95.

After the display of the table, the computer prints the value of the test statistic $\chi^2 = \Sigma(O-E)^2/E$, including the cell-by-cell subtotals. Thus, we see that the value of the test statistic is $\chi^2 = 94.269$. The final item given is the number of degrees of freedom—df = 6.

In order to complete the hypothesis test, we need only determine the critical value and compare it with the value of the test statistic, $\chi^2 = 94.27$. Now, the test is to be performed at the 5% significance level; so $\alpha = 0.05$. The critical value for a chi-square independence test is χ^2_α which, in this case, is $\chi^2_{0.05}$.

PRINTOUT 13.11
Output of INVCDF

```
Chi-Square with 6 DF

 P( X <= x)          x
    0.9500     12.5916
```

Printout 13.9 shows that df = 6. This critical value is not computed for exercising a classical test. The INVCDF command will determine that the critical value is $\chi^2_{0.05} = 12.5916$ as shown in Printout 13.11. Since the value of the test statistic, $\chi^2 = 94.269$, exceeds this critical value, we reject H_0.

Alternatively, since the P-value is practically 0 (it cannot be exactly 0), we reject H_0 at any level of significance. In other words, it appears that marital status and alcohol consumption patterns are statistically dependent. ■

PROBLEM 13.6 Using the contingency table for counts developed in Printout 13.7 to carry out the Chi-square test for independence. ◇

LESSON 14

Descriptive Methods in Regression and Correlation

GENERAL OBJECTIVE

It is frequently of interest to know whether two or more variables are related and, if so, how they are related. For instance, is there a relationship between college GPA and SAT scores? If these variables are related, how are they related? The president of a large corporation knows there is a tendency for sales to increase as advertising expenditures increase. But how strong is that tendency and how can she predict the approximate sales that will result from various advertising expenditures?

Some commonly used methods for examining the relationship between two or more variables and for making predictions are provided by *linear regression and correlation*. Descriptive methods in linear regression and correlation will be addressed in this lesson. Inferential methods in linear regression and correlation will be addressed in the next lesson.

LESSON OUTLINE

14.1 The regression equation
14.2 The coefficient of determination
14.3 Linear correlation

14.1 The regression equation

E-4.2:I-14.2

In this section we will learn how Minitab can be applied to determine the regression line for a set of data points. Recall that the *regression line* is the best-fitting line to a set of data points, best-fitting according to the *least-squares criterion.*

SCATTER DIAGRAMS

The idea behind finding a regression line is based on the assumption that the data points are actually scattered about a straight line. Consequently, before determining the regression line, we need to look at a *scatter diagram* of the data. We should then proceed with the regression only if the data points appear to be scattered about a straight line.

Minitab has the capability of producing scatter diagrams once the data are appropriately stored in the workspace. By examining the way the data point plot in this two-dimensional diagram we can make judgments about whether it is reasonable to assume they are scattered about a straight line. We illustrate in the following example.

EXAMPLE 14.1 *Illustrates creating a scatter diagram*

Table 14.1 (WeissStats CD:Tb14-02) displays data on age and price for a sample of Orion automobiles. The data were obtained from the *Asian Import* edition of the *Auto Trader* magazine. Ages are in years; prices are in hundreds of dollars, rounded to the nearest hundred dollars. The goal is to predict price from age.

TABLE 14.1
Age - price data for Orions

Car	Age (yrs) x	Price ($100s) y
1	5	85
2	4	103
3	6	70
4	5	82
5	5	89
6	5	98
7	6	66
8	6	95
9	2	169
10	7	70
11	7	48

Use Minitab to obtain a scatter diagram of these data.

SOLUTION Begin by entering the data from Table 14.1 into columns named `AGE` and `PRICE` in the usual manner. To obtain a scatter diagram of the age-price data, proceed as follows.

Menu commands:

1 Choose Graph ➤ Plot...
2 Enter `PRICE` in the Y text box for Graph 1
3 Press Tab and enter `AGE` in the X text box
4 Click OK

Session commands:

- Type `Plot 'PRICE'*'AGE'` at the `MTB` > prompt and press Enter

Figure 14.1 shows the resulting output.

FIGURE 14.1 Minitab scatter diagram for the Orion data

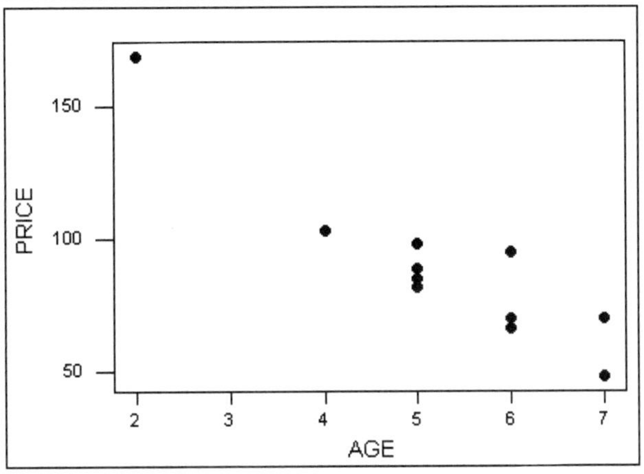

Note that the variable corresponding to the first storage location, in this case, `PRICE`, is plotted on the vertical axis and that the variable corresponding to the second storage location, in this case, `AGE`, is plotted on the horizontal axis. Since `AGE` in this case is the independent variable, we call it the *predictor variable* in regression analysis. Correspondingly, `PRICE`, the dependent variable, is referred to as the *response variable*. Again, the goal is to predict price from age.

You may also elect a character graph by choosing the following sequence of commands: Graphs ➤ Character Graphs ➤ Scatter Plot..., then input the variables as directed. Your output will then resemble Printout 14.1 on the following page.

In general, the data points are plotted as individual plotting symbols (`x` is the default). However, if two or more data points are the same or very close together, then Minitab prints the number of those points instead of the plotting symbol. For example, the `2` plotted above the `5.0` indicates there are two data points in that vicinity that are either the same or very close together. In this case, they are the points $(5, 85)$ and $(5, 89)$.

From either perspective, it does appear that the data points are scattered about a straight line. ■

PRINTOUT 14.1
Minitab scatter plot of Orion data using character graphs

```
Plot

         -
         -    x
      160+
         -
PRICE    -
         -
         -
      120+
         -
         -                        x
         -                                  x         x
         -                                  2
       80+                                  x
         -                                            x         x
         -                                            x
         -
         -                                                      x
          ----+---------+---------+---------+---------+---------+--AGE
             2.0       3.0       4.0       5.0       6.0       7.0
```

Since a straight line model seems justified in this case, it is reasonable to find a regression line for that data. Fortunately, Minitab will simplify this process and we turn to that matter next.

PROBLEM 14.1 The U.S. National Center for Health Statistics collects and publishes data on heights and weights by age and sex. A random sample of 11 males, aged 18–24 years, gave the data in the following table.

Height (in) x	Weight (lb) y
65	175
67	133
71	185
71	163
66	126
75	198
67	153
70	163
71	159
69	151
69	155

Use Minitab to obtain a scatter diagram of the data for males aged 18–24 years using height as a predictor variable and weight as a response variable.Use `HEIGHT` and `WEIGHT` as names for the columns of data. ◇

OBTAINING THE REGRESSION EQUATION

Formula 14.1 in *IS/6e* (Formula 4.1 in *ES/5e*) provides the formulas to determine the regression equation for a set of data points. Minitab has a program that, among other things, will determine the regression equation for us. That program is called REGRESS. We will apply REGRESS to obtain the regression equation for the Orion data.

Before we have Minitab perform a regression analysis on the data, we need to point out that there are four forms for the output. The command BRIEF, followed by an integer between 0 and 3, controls the amount of output for all succeeding regression commands. The larger the integer, the more extensive the output, with BRIEF 2 usually the default and BRIEF 0 being no output at all. For this example, we do not need extensive output. In the menu commands that follow, BRIEF 2 is the default as it is in the session commands so it is necessary to override it. Proceed as follows.

EXAMPLE 14.2 *Illustrates finding a regression equation*

The data on age and price for a sample of Orions are given in Table 14.1. Use Minitab to obtain the regression equation for that data.

SOLUTION Proceed as follows.

Menu commands:

1 Choose Stat ➤ Regression ➤ Regression...
2 Enter `PRICE` in the Response: text box
3 Select the Predictors: text box and enter `AGE`
4 Select the Results:... button and select the second radio button entitled Regression equation... and basic analysis of variance
5 Select [OK] twice

Session commands:

- Type `Regress 'PRICE' 1 'AGE';` at the `MTB >` prompt and press [Enter↵]
- Type `Constant;` at the `SUBC>` prompt and press [Enter↵]
- Type `Brief 1.` at the `SUBC>` prompt and press [Enter↵]

The regression command tells Minitab to perform a regression analysis on the age and price data with `AGE` as the single independent (predictor) variable and `PRICE` the dependent (response) variable with abbreviated output. Printout 14.2 on the following page displays the results. The first item in the output gives the required regression equation:

```
PRICE = 195 - 20.3 AGE
```

In other words, the regression equation is $\hat{y} = 195 - 20.3x$.

PRINTOUT 14.2
Minitab regression output

```
The regression equation is
PRICE = 195 - 20.3 AGE

Predictor        Coef       StDev          T        P
Constant       195.47       15.24      12.83    0.000
AGE           -20.261       2.800      -7.24    0.000

S = 12.58       R-Sq = 85.3%      R-Sq(adj) = 83.7%

Analysis of Variance

Source           DF          SS          MS         F        P
Regression        1      8285.0      8285.0     52.38    0.000
Residual Error    9      1423.5       158.2
Total            10      9708.5
```

After the regression equation, the output provides a table that gives information about the y-intercept, b_0, and slope, b_1, of the regression line. The row labeled `Constant` provides information about b_0 and the row labeled `AGE` provides information about b_1. In particular, the entries `195.47` and `-20.261` under the column headed `Coef` are simply the values of b_0 and b_1, respectively. We will discuss the other aspects of the output in Printout 14.2 in future sections. ■

Now that we have the scatter plot and the equation for the regression line, it is possible to re-plot the graph with the equation superimposed as shown in Figure 14.2. This graph can be constructed using a macro called `%Fitline`. Use the steps on the following page. The graph displays the departure of each sample point from the fitted straight line.

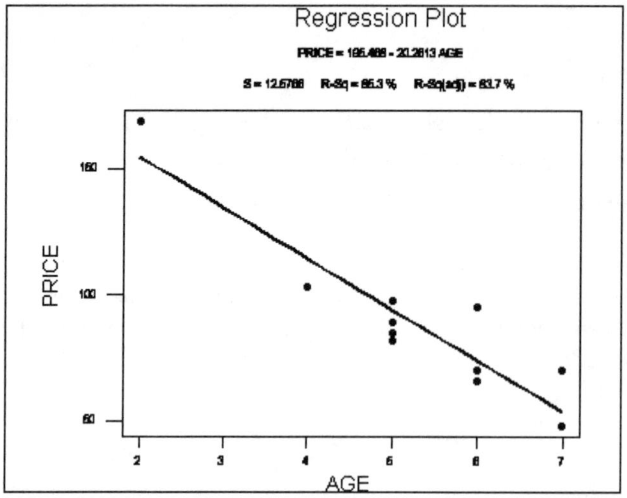

FIGURE 14.2
Minitab scatter diagram for the Orion data with fitted regression line

Menu commands:

1 Choose Stat ➤ Regression ➤ Fitted Line Plot...
2 Enter PRICE in the Response [Y]: text box
3 Enter AGE in the Predictor [X]: text box
4 Click the Linear radio button under Type of Regression Model
5 Select [OK]

Session commands:

- Type %Fitline 'PRICE' 'AGE'

In addition to the graph, you see the equation of the line printed out along with the statistics in the sixth line of Printout 14.2.

PROBLEM 14.2 Refer to Problem 14.1. Use Minitab to obtain the regression equation of the weight-versus-height data, using WEIGHT as the response variable and HEIGHT as the predictor. ◇

14.2 The coefficient of determination

E-4.3:I-14.3

Section 14.3 of *IS/6e* (Section 4.3 of *ES/5e*) discusses the *coefficient of determination, r^2*. The coefficient of determination is a descriptive measure of the utility of the regression line for making predictions. Specifically, r^2 gives the percentage reduction in the total squared error obtained by using the regression equation to predict the observed y-values, instead of just predicting the mean $\overline{y}$.

The coefficient of determination, r^2, is defined in terms of the *total sum of squares, $SST = \Sigma(y - \overline{y})^2$*, and the *error sum of squares, $SSE = \Sigma(y - \hat{y})^2$*. We have

$$r^2 = 1 - \frac{SSE}{SST}$$

Example 14.7 of *IS/6e* (Example 4.7 of *ES/5e*), traces the calculations required to find the coefficient of determination of the age-versus-price data for Orions. As we saw, those calculations are extensive. Minitab will do all of that work for us. In fact, the output of the regression command, which we used to obtain the regression equation, also gives the value of the coefficient of determination.

Look back at Printout 14.2. The second entry in the sixth line of the output displays the coefficient of determination. It appears as R-sq = 85.3%. That is, $r^2 = 0.853$.

PROBLEM 14.3 Refer to Problem 14.2. Use the Minitab output you obtained in that problem to determine the coefficient of determination, r^2. ◇

EXPLAINED VARIATION

There is another way to interpret the coefficient of determination, r^2. Namely, it can be interpreted as the percentage of variation in the observed y-values that is explained by the regression. The justification for this interpretation is given in *IS/6e* and *ES/5e*. We briefly discuss it here.

The deviation, $y - \overline{y}$, of an observed y-value from the mean can be decomposed into two parts—the deviation that is explained by the regression line, $\hat{y} - \overline{y}$, and the remaining unexplained deviation (error), $y - \hat{y}$. Therefore, the total amount of variation (squared deviation) explained by the regression line is $\Sigma(\hat{y} - \overline{y})^2$. This is called the *regression sum of squares*, and is denoted SSR. Hence,

$$SSR = \Sigma(\hat{y} - \overline{y})^2$$

The total variation in the observed y-values is given by the total sum of squares, $SST = \Sigma(y - \overline{y})^2$. Consequently, the percentage of the total variation in the observed y-values that is explained by the regression is SSR/SST. It can be shown that this last quantity is equal to the coefficient of determination, r^2, and it is always true that $SSR + SSE = SST$ (the regression identity).

Minitab can be quite helpful in determining these various sums of squares. To obtain the three sums of squares, we use the table in the output entitled `Analysis of Variance`. Specifically, the values of the three sums of squares can be found in the column headed `SS`. The entries in the column headed `SOURCE` identify the sums of squares. Hence, the first entry in the `SS` column is the regression sum of squares, SSR, the second is the error sum of squares, SSE, and the third is the total sum of squares, SST.

EXAMPLE 14.3 Illustrates computing the coefficient of determination

Use Printout 14.2 to identify the various sums of squares for the price-versus-age data of the Orions. Use the results to interpret the coefficient of determination.

SOLUTION From Printout 14.2, the first entry in the `SS` column is the regression sum of squares, SSR, the second is the error sum of squares, SSE, and the third is the total sum of squares, SST. So, we see that $SSR = 8285.0$, $SSE = 1423.5$, and $SST = 9708.5$. As expected, $SST = SSR + SSE$.

The percentage of the total variation in the observed y-values that is explained by the regression is

$$\frac{SSR}{SST} = \frac{8285.0}{9708.5} = 0.853$$

or 85.3%. In other words, about 85% of SST is equal to the regression sum of squares SSR. This is the sense in which we interpret the phrase "explained by regression." As noted earlier, this quantity is equal to the coefficient of determination, r^2. We will encounter this same result in the context of correlation in the next section. ■

PROBLEM 14.4 Refer to Problem 14.2 and Problem 14.3. Use the Minitab output you obtained in Problem 14.2 to find the regression sum of squares SSR, the error sum of squares, SSE, and the total sum of squares, SST. Determine the coefficient of determination from these sums of squares. ◇

14.3 Linear correlation

E-4.4:I-14.4

It is often of interest to know whether there is an association or *correlation* between two variables. For example, is there a positive correlation between advertising expenditures and sales? Or, are IQ and alcohol consumption uncorrelated?

One of the most common measures of the correlation between two variables is the *linear correlation coefficient,* r. The linear correlation coefficient is a single number that can be used to describe the strength of the *linear* (straight-line) relationship between two variables.

The linear correlation coefficient, r, is always between -1 and 1. Positive values of r suggest that the variables x and y are *positively linearly correlated,* meaning that y tends to increase linearly as x increases, with the tendency being greater the closer that r is to 1. Negative values of r suggest that the variables are *negatively linearly correlated,* meaning that y tends to decrease linearly as x increases, with the tendency being greater the closer that r is to -1. Values of r close to 0 suggest that the variables are *linearly uncorrelated,* meaning that there is neither a tendency for y to increase as x increases nor a tendency for y to decrease as x increases—in a *linear* fashion.

The linear correlation coefficient, r, for n data points can be computed using the formula

$$r = \frac{n(\Sigma xy) - (\Sigma x)(\Sigma y)}{\sqrt{n(\Sigma x^2) - (\Sigma x)^2}\sqrt{n(\Sigma y^2) - (\Sigma y)^2}}$$

Minitab has a routine that will compute r for us. The next example illustrates this.

EXAMPLE 14.4 Illustrates correlation

The data on age and price for a sample of 11 Orions are displayed in the second and third columns of Table 14.1. Use Minitab to determine the linear correlation coefficient of the data.

SOLUTION Recall that we have previously stored the age and price data in columns named `AGE` and `PRICE`. Proceed as follows.

Menu commands:

1 Choose Stat ➤ Basic Statistics ➤ Correlation...

2 Enter `AGE` and `PRICE`, in either order, in the Variables: text box

3 Select [OK]

Session commands:

- Type CORRELATION 'AGE' 'PRICE'

Printout 14.3 displays the result.

PRINTOUT 14.3
Minitab correlation

```
Pearson correlation of AGE and PRICE = -0.924
P-Value = 0.000
```

From the printout, we conclude that the linear correlation coefficient for the age and price data is −0.924; that is, $r = -0.924$. Evidently, there is a strong negative linear correlation between age and price of Orions. This means that as the age of a Orion increases, the price tends to decrease in a linear fashion. Incidentally, it is no coincidence that the coefficient of determination is denoted r^2, for the numerical value of the correlation coefficient, $|r|$, is indeed $\sqrt{r^2}$. The printed *P*-value (not available in earlier releases) will be discussed in the next section. ■

PROBLEM 14.5 Use the weight-versus-height data for males 18–24 years old in Problem 14.1 to find the linear correlation coefficient of HEIGHT and WEIGHT. ◇

LESSON 15

Inferential Methods in Regression and Correlation

GENERAL OBJECTIVE

We often need to know whether a relationship exists between two variables, and if so, what that relationship is. For example, is there a relationship between SAT score and college grade-point average? If these variables are related, how are they related? Linear regression analysis and correlation analysis provide some of the most widely used procedures for examining the relationship between two variables. In this lesson we will learn how Minitab can be applied to perform linear regression and correlation analyses.

LESSON OUTLINE

15.1 The regression model: analysis of residuals

E-14.1:I-15.1

Up to now we have only examined *descriptive* methods in linear regression and correlation. In order to make *inferences* in linear regression and correlation, it is necessary for the variables under consideration to satisfy certain conditions. We list them below.

Assumptions for regression inferences

1. *Population regression line:* There are constants β_0 and β_1 such that for each value of the predictor variable, x, the conditional mean of the response variable is $\beta_0 + \beta_1 x$. We refer to the straight line $y = \beta_0 + \beta_1 x$ as the *population regression line* and to its equation as the *population regression equation.*
2. *Equal standard deviations:* The conditional standard deviations of the response variable are the same for all values of the predictor variable. We denote this common standard deviation by σ.
3. *Normality:* For each value of the predictor variable, the conditional distribution of the response variable is a normal distribution.
4. *Independent observations:* The observations of the response variable are independent of one another.

In other words, Assumptions (1) –(3) require that there exist constants β_0, β_1, and σ so that for each value of x, the predictor variable, the conditional distribution of the response variable, y, is a normal distribution having mean $\beta_0 + \beta_1 x$ and standard deviation σ. These assumptions are often referred to as the *regression model.*

THE STANDARD ERROR OF THE ESTIMATE

Let us suppose that we are considering two variables, x and y, satisfying the regression model. The parameters β_0, β_1, and σ, are almost always unknown and, hence, must be estimated from the sample data. The statistic used to estimate σ is called the *standard error of the estimate* or the *residual standard deviation* and is defined as follows:

$$s_e = \sqrt{\frac{SSE}{n-2}}$$

where, as we know, $SSE = \Sigma(y - \hat{y})^2$.

In Section 14.1, we learned that Minitab can be used to obtain the (sample) regression equation for a set of data points. The output resulting from the application of the Regression command displays the value of the standard error of the estimate, s_e, (labeled s on the output) as well as the regression equation and several other statistics. The standard error of the estimate plays an important role in the regression model where it is assumed that the variance or standard deviation is the same regardless of the value of the predictor variable.

EXAMPLE 15.1 *Illustrates finding the standard error of the estimate*

We have duplicated Printout 14.2, the Minitab regression output applied to the data on age and price for a sample of 11 Orions given in Table 14.1, in Printout 15.1.

PRINTOUT 15.1 Minitab regression output

```
The regression equation is
PRICE = 195 - 20.3 AGE

Predictor        Coef        Stdev     t-ratio        p
Constant       195.47        15.24       12.83    0.000
AGE           -20.261        2.800       -7.24    0.000

s = 12.58        R-sq = 85.3%     R-sq(adj) = 83.7%

Analysis of Variance

SOURCE       DF          SS          MS         F        p
Regression    1      8285.0      8285.0     52.38    0.000
Error         9      1423.5       158.2
Total        10      9708.5
```

Use the output in Printout 15.1 to determine the standard error of the estimate for the sample of 11 Orions.

SOLUTION The standard error of the estimate, s_e, is the first entry in the sixth line of the output: `s = 12.58`. Thus, for the Orion data, $s_e = 12.58$. This is the estimate of the constant standard deviation in the regression model. ■

PROBLEM 15.1 Refer to the Minitab output you obtained in Problem 14.2

a) What is s_e for the height and weight data?
b) Presuming that the variables height, x, and weight, y, for males 18–24 years old satisfy the assumptions for regression inferences, what is your estimate of the common population standard deviation, σ, of weights for males 18–24 years old of any given height? ◇

ANALYSIS OF RESIDUALS

We can also use the regression routine to obtain residuals and then plot the results. We can then examine the residual plot for linearity. We illustrate this in the next example.

EXAMPLE 15.2 *Illustrates residuals and residual plots*

Using the data of the preceding example, obtain the residuals, residual plot, and a normal probability plot of the residuals for the age and price data of the Orions.

SOLUTION With the data still stored in columns named PRICE and AGE, proceed as follows.

Menu commands:

1 Choose Stat ➤ Regression ➤ Regression...
2 Enter PRICE in the Response: text box
3 Select the Predictors: text box and enter AGE
4 Select the Graphs... button
5 Select the Regular option button under Residual for Plots: options
6 Select the Normal plot of residuals check box
7 Select the Residuals versus the variables: text box and enter AGE
8 Click [Results] and select the second radio button
9 Click [OK] twice

In Figure 15.1, we show the resulting graph output for the residual plot.

FIGURE 15.1
Minitab residual plot

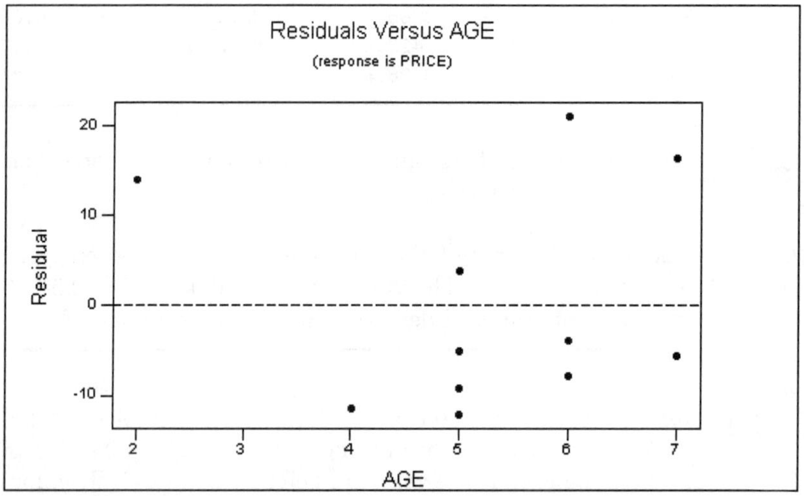

The sequence of steps shown in the next printout constitute the session commands that will achieve the same results.

PRINTOUT 15.2
Minitab session commands

```
MTB > Regress 'PRICE' 1 'AGE';
SUBC>  GNormalplot;
SUBC>  GVars 'AGE';
SUBC>  RType 1;
SUBC>   Constant;
SUBC>   Brief 1.
```

The residuals fall roughly in a horizontal band centered about 0 with no clear pattern on the AGE axis so we are satisfied with the regression assumptions.

In Figure 15.2, we show the resulting probability plot of the residuals converted to normal scores.

FIGURE 15.2
Minitab normal probability plot of residuals

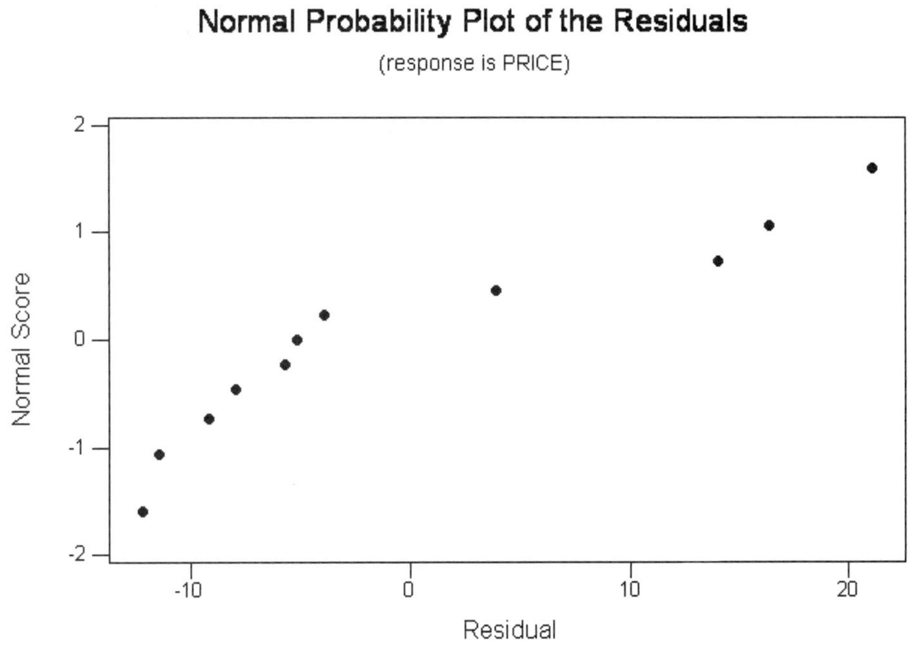

We may say that roughly the points seem to scatter around a straight line so that there is no obvious violation of the regression assumptions. Of course you would not expect to get a perfect straight line even if the model were correct since there is a random factor in the sample values. ■

PROBLEM 15.2 Refer to Problem 14.2. Using Minitab, obtain the residuals, residual plot, and a normal probability plot of the residuals for the regression of weight on height. ◇

15.2 Inferences for the slope of the population regression line

E-14.2:I-15.2

In this section, we will learn how Minitab can be used to perform the inferential procedures discussed in Section 15.2 of *IS/6e* (Section 14.2 of *ES/5e*). That is, we will examine inferences concerning β_1—the slope of the population regression line. The inference centers about the fact that if the slope of the regression line were 0, then one variable would not be very useful in predicting the other.

HYPOTHESIS TESTS FOR THE SLOPE OF THE POPULATION REGRESSION LINE

A step-by-step method for performing a hypothesis test to decide whether the slope, β_1, of a population regression line is not zero is given in Procedure 15.1 in *IS/6e* (Procedure 14.1 in *ES/5e*). Alternatively, we can use Minitab to carry out such a hypothesis test. In fact, Minitab's regression output contains all the information that we need. To illustrate, we return to the Orion example previously treated.

EXAMPLE 15.3 *Illustrates the t-test of utility of regression*

In Section 14.2 we applied the REGRESS command to the data on age and price for a sample of 11 Orions shown in Table 14.1. The resulting computer output is repeated here in Printout 15.3.

PRINTOUT 15.3 Minitab regression output

```
The regression equation is
PRICE = 195 - 20.3 AGE

Predictor        Coef        Stdev    t-ratio        p
Constant       195.47        15.24      12.83    0.000
AGE           -20.261        2.800      -7.24    0.000

s = 12.58        R-sq = 85.3%     R-sq(adj) = 83.7%

Analysis of Variance

SOURCE       DF          SS          MS          F        p
Regression    1      8285.0      8285.0      52.38    0.000
Error         9      1423.5       158.2
Total        10      9708.5
```

Perform the test

$$H_0\text{: } \beta_1 = 0 \text{ (age is not useful for predicting price)}$$
$$H_a\text{: } \beta_1 \neq 0 \text{ (age is useful for predicting price)}$$

at the 5% significance level to decide whether or not the data provide sufficient evidence that age is useful as a predictor of price for Orions. Assume that the variables, age, x, and price, y, satisfy the assumptions for regression inferences (page 201). Use the Minitab output.

SOLUTION Look at the fifth line of the computer output in Printout 15.3, the line labeled **AGE**. The second entry in that line, which is under the column headed **Coef**, gives the slope, b_1, of the sample regression line; hence, $b_1 = -20.261$. The third entry in that line, which is under the column headed **Stdev**, shows the estimated standard deviation of b_1, $s_e/\sqrt{S_{xx}}$. So, $s_e/\sqrt{S_{xx}} = 2.800$. The fourth entry in that line, which is under the column headed `t-ratio`, displays the value of the test statistic

$$t = \frac{b_1}{s_e/\sqrt{S_{xx}}}$$

Thus, we see that $t = -7.24$.

The final entry of the line labeled `AGE`, under the column headed `p`, gives the P-value for the hypothesis test. This is really the only quantity that we need in order to decide whether the null hypothesis should be rejected. Recall that the P-value is equal to the smallest significance level at which the null hypothesis can be rejected. Therefore, if the P-value is less than or equal to the specified significance level, then we reject H_0; otherwise, we do not reject H_0.

From Printout 15.3, we see that the P-value is 0.000 (to three decimal places). Since this is less than the specified significance level of $\alpha = 0.05$, we reject H_0. In other words, the data provide sufficient evidence to conclude that the slope of the population regression line is not zero and, hence, that age is useful as a predictor of price for Orions. ■

PROBLEM 15.3 In Problem 14.2, you performed a regression of weight on height for a sample of males 18–24 years old. Use the output you obtained, to decide whether the slope, β_1, of the population regression line is not zero. Perform the hypothesis test at the 5% significance level. ◇

CONFIDENCE INTERVALS FOR THE SLOPE OF THE POPULATION REGRESSION LINE

We can also employ Minitab to help us find a *confidence interval* for the slope, β_1, of the population regression line following the two steps given in Procedure 15.2 in *IS/6e* (Procedure 14.2 in *ES/5e*). Again, we will use the age and price data for Orions to illustrate the steps, repeated here.

EXAMPLE 15.4 ***Illustrates how to use Minitab in conjunction with Procedure 15.2 of IS/6e (Procedure 14.2 of ES/5e)***

Use the sample data in Table 14.1 to find a 95% confidence interval for the slope, β_1, of the population regression line relating age, x, and price, y, of Orions.

SOLUTION We will employ Minitab in conjunction with Procedure 15.2 of *IS/6e* (Procedure 14.2 of *ES/5e*).

STEP 1 *For a confidence level of 1 − α, find $t_{\alpha/2}$ for df = n − 2.*

For a 95% confidence interval, $\alpha = 0.05$. Since $n = 11$, df $= 11 - 2 = 9$. Thus, we need to find $t_{\alpha/2} = t_{0.05/2} = t_{0.025}$ for df $= 9$. We can do this with Minitab using the INVCDF command. Printout 15.4 on the following page shows the result.

PRINTOUT 15.4
Output of INVCDF

```
Student's t distribution with 9 d.f.

  P( X <= x)          x
     0.9750      2.2622
```

The printout shows that $t_{0.025} = 2.2622$.

STEP 2 *The endpoints of the confidence interval are*

$$b_1 \pm t_{\alpha/2} \cdot s_e/\sqrt{\Sigma x^2 - (\Sigma x)^2/n}$$

From Step 1, $t_{\alpha/2} = 2.2622$. We can find the remaining quantities required for the evaluation of the confidence interval from the regression output. Look back at Printout 15.2. The fifth line of the output begins with `AGE`. The second entry in that line, which is in the column headed `Coef`, gives b_1. So we see that $b_1 = -20.261$. The third entry in that line, which is in the column headed `Stdev`, gives the estimated standard deviation of b_1; that is, $s_e/\sqrt{\Sigma x^2 - (\Sigma x)^2/n}$. So,

$$s_e/\sqrt{\Sigma x^2 - (\Sigma x)^2/n} = 2.800$$

Consequently, the endpoints of the confidence interval are

$$-20.261 \pm 2.2622 \cdot 2.800$$

or

$$-26.5952 \quad \text{to} \quad -13.9268$$

We can be 95% confident that β_1 is between -26.5952 and -13.9268. In other words, we can be 95% confident that the yearly drop in mean price for Orions is somewhere between \$1392.68 and \$2659.52.

Of course, for a different confidence level, you have only to adjust $t_{\alpha/2}$ in the formula at Step 2. For example, to find a 90% confidence interval, you would use Minitab to discover that $t_{0.05} = 1.8331$. You may verify that the 90% confidence interval now runs from -25.3934 to -15.1286. ■

PROBLEM 15.4 Refer to the output of the REGRESS command that you obtained in Problem 14.2. Use that output along with the techniques of Example 15.4 to find a 95% confidence interval for the slope, β_1, of the population regression line relating height, x, and weight, y, of males 18–24 years old. ◇

15.3 Estimation and prediction

E-14.3:I-15.3

In this section, we will learn how to use Minitab to help make two important types of inferences. One is to estimate the conditional mean of the population of the response variable corresponding to a particular value of the predictor variable. The other is to predict the value of the response variable for a particular value of the predictor variable. There is a subtle difference between the two intervals. In the case of the conditional mean, the confidence interval is an interval estimate of the parameter as before. In the case of a prediction interval, we are trying to estimate the actual value of the response variable, not just its mean value. This would appear to be a more difficult task since we are dealing with the value itself, not just the mean of all possible values.

We will employ the Orion example yet again to illustrate the pertinent ideas. Recall from the data in Table 14.1 that ages are in years; prices are in hundreds of dollars, rounded to the nearest hundred. We assume these data are entered and appropriately named in the workspace as in the previous lesson.

Also recall that the sample regression line for the data in Table 14.1 is determined by the equation $\hat{y} = 195.47 - 20.261x$, using the values given under `Coef`. Thus, our estimate for the mean price of all three-year-old Orions is

$$\hat{y} = 195.47 - 20.261 \cdot 3 = 134.69$$

or \$13,469. [Note that the estimate for the mean price of all three-year-old Orions is the same as the predicted price of a randomly selected three-year-old Orion. Both are obtained by substituting $x = 3$ into the sample regression equation.]

The estimate of \$13,469 for the mean price of all three-year-old Orions is a point estimate. As we know, it would be more informative if we had some idea of the accuracy of that point estimate. In other words, it would be better to provide a confidence-interval estimate for the mean price of all three-year-old Orions. We will see how Minitab can be used to find this confidence interval using a set of subcommands available for the basic REGRESS command.

CONFIDENCE INTERVALS FOR CONDITIONAL MEANS

Minitab has regression options that may be used directly for estimation and prediction. We illustrate in the next example.

EXAMPLE 15.5 Illustrates confidence interval estimation

Data on age and price for a sample of 11 Orions are found in Table 14.1. Apply Minitab to that data to obtain a 95% confidence interval for the mean price of all three-year-old Orions.

SOLUTION We assume that the data are located in columns named `AGE` and `PRICE`. Proceed as follows.

Menu commands:

1 Choose Stat ➤ Regression ➤ Regression...

2 Enter `PRICE` as the response variable and `AGE` as the predictor variable as before

3 Select the Options... button

4 Select the Prediction intervals for new observations: text box and type 3

5 Select the Confidence level: text box and type 95 (if necessary)

6 Select OK twice

Session commands: Type `REGRESS 'PRICE' 1 'AGE';` at the `MTB >` prompt and press Enter

- Type `Constant;` at the `SUBC>` prompt and press Enter
- Type `Predict 3.` at the `SUBC>` prompt and press Enter

Printout 15.5 shows the last six lines of the resulting output.

PRINTOUT 15.5
Minitab output for confidence intervals

```
Predicted Values for New Observations

New Obs      Fit      SE Fit          95.0% CI                95.0% PI
1         134.68        7.41    ( 117.93,  151.44)   ( 101.67,  167.70)

Values of Predictors for New Observations

New Obs        AGE
1             3.00
```

Under the entry headed `Fit` you will find the point estimate, $\hat{y}_p = 134.68$ ($13,468), for the mean price of all three-year-old Orions or for the price of a randomly selected three-year-old Orion. The item under the heading `SE Fit`, displays the estimated standard deviation of the random variable $\hat{y}_p = b_0 + b_1 x_p$, which, for $x_p = 3$, is

$$s_e\sqrt{\frac{1}{n} + \frac{(x_p - \Sigma x/n)^2}{S_{xx}}} = 7.41$$

Under the entry headed `95% C.I.`, you will find the required confidence interval. Hence, a 95% confidence interval for the mean price of all three-year-old Orions is from 117.93 to 151.44. We can be 95% confident that the mean price of all three-year-old Orions is somewhere between $11,793 and $15,144. ■

PROBLEM 15.5 The height and weight data for a sample of males 18–24 years old may be found in the table for Problem 14.1. Use Minitab to determine a 95% confidence interval for the mean weight of all males 18–24 years old who are 70 inches tall. ◇

PREDICTION INTERVALS

One of the main reasons for determining the regression equation for a sample of data points is to use it for making predictions. Procedure 15.4 located in *IS/6e*

(Procedure 14.4 in *ES/5e*) gives a step-by-step method for obtaining a prediction interval for the value of the response variable corresponding to a particular value of the predictor variable. You may already have noticed that this is part of the output for confidence intervals. Consider the following example.

EXAMPLE 15.6 *Illustrates prediction intervals*

The age and price data for a sample of Orions are given in Table 14.1. Apply Minitab to that data to obtain both a 95% prediction interval and a 90% prediction interval for the price of a randomly selected three-year-old Orion.

SOLUTION In Example 15.5 we found a 95% confidence interval for the mean price of all three-year-old Orions. The output given in Printout 15.5 also provides the desired 95% prediction interval under the heading `95% P.I.`. So, a 95% prediction interval for the price of a randomly selected three-year-old Orion is from 101.67 to 167.70. We can be 95% certain that the price of a randomly selected three-year-old Orion will be somewhere between $10,167 and $16,770.

Contrast this with the confidence interval for the mean of all such Orions. You see that the prediction interval is wider, allowing for more possibilities than the confidence interval. In other words, we are less sure of what the price of a single three-hear-old Orion would be compared to the average price of all three-year-old Orions. ■

PROBLEM 15.6 Refer to Problem 15.5. Use the output you obtained in Problem 15.5 to find both a 95% prediction interval for the weight of a randomly selected male 18–24 years old who is 70 inches tall. ◇

15.4 Inferences in correlation

E-14.4:I-15.4

In the previous section we learned that we can decide whether two variables, x and y, are linearly related by performing a hypothesis test for the slope, β_1, of the population regression line. Alternatively, we can perform a hypothesis test for the *population linear correlation coefficient,* ρ (rho).

The population linear correlation coefficient, ρ, measures the linear correlation between the population of *all* data points in the same way that the (sample) linear correlation coefficient, r, measures the linear correlation between a sample of data points. Thus, it is ρ that actually describes the linear correlation between two variables.

Procedure 15.5 in *IS/6e* (Procedure 14.5, in *ES/5e*) is a step-by-step method for performing a hypothesis test with null hypothesis $H_0 : \rho = 0$ (that is, the variables are linearly uncorrelated). Rejection of the null hypothesis indicates that the variables under consideration are linearly related, as in the next example. We discovered in Section 14.3 that Minitab can be used to compute the value of r and automatically produces a P-value for this test. We can expand on this as the next example shows. The test statistic for the hypothesis test is a function of the (sample) correlation coefficient, r. This will carry out Procedure 15.5, as the next example shows.

EXAMPLE 15.7 Illustrates how Minitab can be used in conjunction with Procedure 15.5 in IS/6e (Procedure 14.5 in ES/5e)

Consider once more the age and price data for Orions. Do the data provide sufficient evidence to conclude that the variables age, x, and price, y, for Orions are *negatively* linearly correlated? Use $\alpha = 0.05$.

SOLUTION In Example 14.5, we used Minitab's CORRELATION command to obtain the linear correlation coefficient, r, of the age and price data. We repeat the output of Printout 14.3 here as Printout 15.6.

PRINTOUT 15.6
Minitab correlation

```
Correlation of PRICE and AGE = -0.924, P-Value = 0.000
```

With a P-value of 0.000, we surely reject the hypothesis that the variables, age and price, for Orions are linearly uncorrelated. However, Minitab reports the P-value for a two-tailed test so that it will only allow us to conclude that the true correlation is something other than 0. As posed, the question suggests a left-tailed test, which, because of the extreme value of the sample correlation is clearly 0 also.

However, we can carry out a left-tailed test using the critical value approach as shown in the following steps.

STEP 1 *We have $n = 11$, so that df $= 9$. Also $\alpha = 0.05$. Using Minitab's* INVCDF command, we found that $-t_{0.05} = -1.8331$. This is the critical value for a left-tailed test.

STEP 2 *In Printout 15.8 , we see that $r = -0.924$. Therefore, the value of the test statistic is*

$$t = \frac{r}{\sqrt{\dfrac{1-r^2}{n-2}}} = \frac{-0.924}{\sqrt{\dfrac{1-(-0.924)^2}{11-2}}} = -7.249$$

STEP 3 *Since the value of the test statistic is less than the critical value, we reject H_0.*

We elect to conclude that the variables are negatively linearly correlated. In other words, the price of an Orion tends to decrease linearly as its age increases (at least for Orions between two and seven years old). ■

PROBLEM 15.7 Consider again the height and weight data for a sample of males 18–24 years old treated in Problem 14.1. Do the data provide sufficient evidence to conclude that the variables height and weight for males 18–24 years old are *positively* linearly correlated? Perform the test at the 5% significance level. ◇

15.5 Testing for normality

I-15.5

In Section 15.5 of *IS/6e* a point is made that our graphical procedures for assessing normality are very subjective and require experience in data analysis. A procedure is then outlined for carrying out a more objective hypothesis test for normality based on computing the linear correlation coefficient between the sample data and the normal scores of the data and then consulting Table XIV. The details are summarized in Procedure 15.6 in *IS/6e* . Minitab may be used to carry out this procedure in a simpler way as the next example illustrates.

EXAMPLE 15.8 *Illustrates a correlation test of normality*

The Internal Revenue Service publishes data on federal individual income tax returns in *Statistics of Income, Individual Income Tax Returns.* A sample of 12 returns yielded the adjusted gross incomes, in thousands of dollars, shown in the following table (WeissStats CD:Tb15-05).

TABLE 15.1 Adjusted gross incomes ($1000)

9.7	93.1	33.0	21.2
81.4	51.1	43.5	10.6
12.8	7.8	18.1	12.7

Use Minitab to test the data for normality.

SOLUTION Enter the data into a column named AGI in the usual manner.

Menu commands:

1 Choose Stat ➤ Basic Statistics ➤ Normality Test...
2 Select the Variable: text box and enter AGI
3 Select the Ryan-Joiner option button under Tests for Normality
4 Select [OK]

Session commands:

- Type `%NormPlot 'AGI';` at the MTB > prompt and press [Enter↵]
- Type `swtest.` at the SUBC> prompt and press [Enter↵]

See Figure 15.3 on the following page for the result.

FIGURE 15.3
Minitab normal probability plot for AGI data

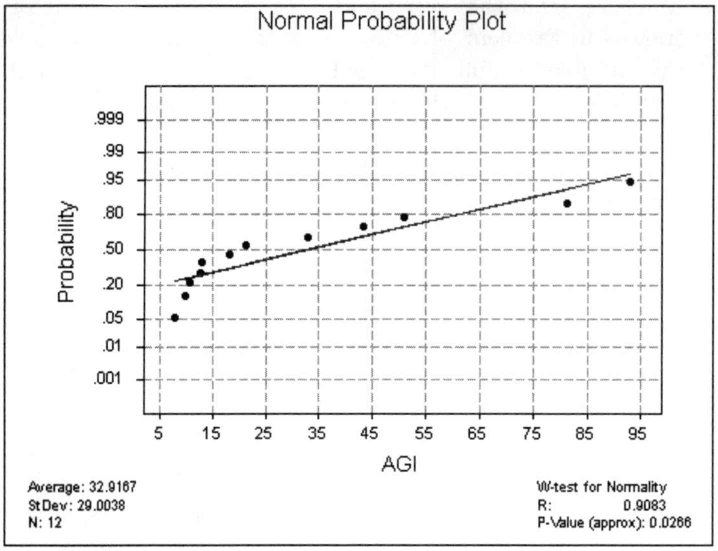

From the figure, you see that the correlation between **NSCORES** and **AGI** is 0.9083. This produces a P-value of 0.0266 and so you would reject the null hypothesis at the 5% level of significance. That is, you would conclude that there is sufficient evidence to conclude that adjusted gross incomes are *NOT* normally distributed since the null hypothesis is that the variable is so distributed. ■

PROBLEM 15.8 The following table constitutes a random sample of 12 of subscriptions to law periodicals.

$30	46	44	47
42	38	62	55
52	48	43	54

Store the data in a column named **SUBS**. Use Minitab to carry out a test of normality using 5% as the level of significance.

LESSON 16

Analysis of Variance (ANOVA)

GENERAL OBJECTIVE

In Lesson 10 we saw how Minitab can be used to draw inferences concerning the means of two populations. Now we will examine how Minitab can be applied to draw inferences concerning the means of more than two populations. The statistical procedures for comparing the means of more than two populations are referred to as *analysis of variance*, or *ANOVA*.

LESSON OUTLINE

16.1 The F-distribution

16.2 One-way analysis of variance

16.3 Multiple comparisons

16.4 The Kruskal-Wallis test

16.1 The F-distribution

E-13.1:I-16.1

The analysis-of-variance procedures utilize a class of continuous probability distributions called ***F*-*distributions,*** named in honor of Ronald Fisher (1890–1962). We will now study how Minitab can be used to find percentile points, F_{α}, for this distribution. These are required by analysis of variance.

To perform an analysis of variance, we need to be able to find, for a given F-curve, the F-value having a specified area to its right. This can be accomplished using Minitab's INVCDF command in conjunction with the subcommand called F, with a specification of degrees of freedom, a pair of numbers in each case. We explain the details in the following example.

EXAMPLE 16.1 *Illustrates finding F percentiles*

For an F-curve with df = (4, 12), find $F_{0.05}$. That is, find the F-value with area 0.05 to its right.

SOLUTION The command INVCDF with the subcommand F computes the F-value with a specified area to its *left.* We want to find the F-value having area 0.05 to its *right* for an F-curve with df = (4, 12). Since the total area under an F-curve is 1, this is equivalent to finding the F-value having area 0.95 to its *left* for an F-curve with df = (4, 12). Proceed as follows.

Menu commands:

1 Choose Calc ➤ Probability Distributions ➤ F...
2 Select the Inverse cumulative probability radio button
3 Select the Numerator degrees of freedom: text box and type 4
4 Select the Denominator degrees of freedom: text box and type 12
5 Select the Input COnstant: radio button
6 Press [Tab] and type 0.95 in the text box
7 Select [OK]

Session commands:

- Type INVCDF 0.95; at the MTB > prompt and press [Enter]
- Type F 4 12. at the SUBC> prompt and press [Enter]

Printout 16.1 displays the output.

PRINTOUT 16.1 Output of the F subcommand to INVCDF

```
F distribution with 4 DF in numerator and 12 DF in denominator

 P( X <= x)          x
    0.9500      3.2592
```

The output shows that for an F-curve with df = (4, 12), the F-value with area 0.95 to its left is 3.2592. Thus, for an F-curve with df = (4, 12), $F_{0.05} = 3.2592$. ■

PROBLEM 16.1 For an F-curve with df = (12, 5), find
a) $F_{0.05}$ b) $F_{0.01}$ c) $F_{0.025}$ ◇

16.2 One-way analysis of variance

E-13.3:I-16.3

We have already mentioned that analysis of variance (ANOVA) is an inferential procedure that is used to compare the means of several populations. The reason for the word "variance" in "analysis of variance" is that the procedure for comparing the means involves analyzing the variation in the sample data.

In this section we will examine *one-way analysis of variance*. It is called *one-way* analysis of variance because each piece of data is classified in one way; namely, according to the population from which it was sampled. As in the pooled-t procedure, we make the following assumptions:

KEY FACT 16.1 Assumptions for one-way ANOVA

1. ***Independent samples:*** **The samples taken from the populations under consideration are independent of one another.**
2. ***Normal populations:*** **The populations under consideration are normally distributed.**
3. ***Equal standard deviations:*** **The standard deviations of the populations under consideration are equal.**

Assumption 1, the independent-samples assumption, is often referred to as the assumption of a *completely randomized design.* Procedure 16.1 in *IS/6e* (Procedure 13.1 in *ES/5e*) provides a step-by-step method for performing a one-way ANOVA test. Alternatively, we can apply Minitab to carry out such a hypothesis test. The appropriate session command is called AOVONEWAY and the menu command is called ANOVA. The next example explains how these commands are used.

EXAMPLE 16.2 Illustrates ANOVA in Minitab

The U.S. Energy Information Administration gathers data on residential energy consumption and expenditures. A researcher wants to know if there is a difference in mean annual energy consumption among the four regions of the United States. Independent random samples of households in the four U.S. regions yield the data on last year's energy consumption given in Table 16.1 on the following page (WeissStats CD:Tb16-03). Use Minitab to perform the hypothesis test at the 5% significance level.

TABLE 16.1 Samples and their means of last year's energy consumption for households in the four U.S. regions

Northeast	Midwest	South	West	
15	17	11	10	
10	12	7	12	
13	18	9	8	
14	13	13	7	
13	15		9	
	12			
13.0	14.5	10.0	9.2	⟵ *Means*

SOLUTION Let μ_1, μ_2, μ_3, and μ_4 denote last year's mean energy consumption for households in the Northeast, Midwest, South, and West, respectively. Then the null and alternative hypotheses are

H_0: $\mu_1 = \mu_2 = \mu_3 = \mu_4$ (mean energy consumption are equal)

H_a: Not all the means are equal.

To employ Minitab, first enter the four data sets of Table 16.1 into columns named `NRTHEAST`, `MIDWEST`, `SOUTH` and `WEST` in the usual manner. Proceed as follows.

Menu commands:

1 Choose Stat ➤ ANOVA ➤ Oneway [Unstacked]...

2 Enter `NRTHEAST MIDWEST SOUTH WEST` in the Responses [in separate columns]: text box

3 Select [OK]

Printout 16.2 displays the resulting output.

PRINTOUT 16.2 Minitab output for one-way analysis of variance

```
One-Way Analysis of Variance

Analysis of Variance
Source      DF        SS        MS        F        p
Factor       3     97.50     32.50     6.32    0.005
Error       16     82.30      5.14
Total       19    179.80
                                     Individual 95% CIs For Mean
                                     Based on Pooled StDev
 Level       N      Mean     StDev   -------+---------+---------+---------
NRTHEAST     5    13.000     1.871                 (------*-------)
MIDWEST      6    14.500     2.588                       (-----*------)
SOUTH        4    10.000     2.582     (-------*-------)
WEST         5     9.200     1.924   (-------*------)
                                     -------+---------+---------+---------
Pooled StDev =     2.268                   9.0      12.0      15.0
```

The following session commands will achieve the same result.

Session commands:

- Type AOVO 'NRTHEAST' 'MIDWEST' 'SOUTH' 'WEST' at the MTB > prompt and press Enter

As you can see, the output contains a lot of information. Let us, for the moment, concentrate on the items of the output required to perform the hypothesis test. The first part of the computer output gives a one-way ANOVA table. This is Minitab's version of the ANOVA table in Table 16.5 in *IS/6e* (Table 13.5 in *ES/5e*). Note that Minitab uses the terminology "Factor" instead of "Treatment."

To the right of the one-way ANOVA table in Printout 16.2, under the column headed p is the P-value. As we know, this is really the only item that is required to make the decision concerning the hypothesis test. Since the P-value is 0.005, which is less than the specified significance level of $\alpha = 0.05$, we reject H_0. In other words, the data provide sufficient evidence to conclude that last year's mean energy consumption for households in the four U.S. regions are not all the same.

We may also use Minitab to assist us with carrying out the various steps of Procedure 16.1 of *IS/6e* (Procedure 13.1 of *ES/5e*). To obtain the critical value for the test we proceed as follows. The two numbers of degrees of freedom for the appropriate F-curve are displayed as the first two entries in the column headed DF. Thus, df = (3, 16).

We are to perform the hypothesis test at the 5% significance level; so $\alpha = 0.05$. Consequently, the critical value is $F_{0.05}$ for an F-curve with df = (3, 16). We can use Minitab to compute this percentile point just as we did earlier in this section. See Printout 16.3 for the result.

PRINTOUT 16.3
Critical value for test

```
F distribution with 3 d.f. in numerator and 16 d.f. in denominator

 P( X <= x)          x
    0.9500      3.2389
```

Thus, the critical value is $F_{0.05} = 3.2389$. To complete the hypothesis test, we only need to compare the value of the test statistic, F, to the critical value. The value of F is displayed in the column headed F in Printout 16.2. Thus, $F = 6.32$.

Observe that the value of the test statistic, $F = 6.32$, exceeds the critical value, $F_{0.05} = 3.2389$. Consequently, since the hypothesis test is right-tailed, we reject H_0. In other words, the data indicate that last year's means of energy consumption for households in the four U.S. regions are not all the same. ■

OTHER ITEMS IN THE OUTPUT

We will now examine some of the other items in the output shown in Printout 16.2. Recall the fundamental identity in one-way analysis of variance which is identified as *one-way ANOVA identity:*

$$SST = SSTR + SSE$$

This identity shows that, indeed, the total variation among all the sample data can be partitioned into a component representing variation among sample means and a component representing variation within samples. The output of the AOVONEWAY and ANOVA commands also displays the values of the three sums of squares—SST, $SSTR$, and SSE.

To illustrate, examine Printout 16.2 once more. The three sums of squares are printed in the third column, which is headed `SS` (for sums of squares). Hence, we see that for the energy-consumption data, $SSTR = 97.50$, $SSE = 82.30$, and $SST = 179.80$. Note that for these sums of squares

$$179.80 = 97.50 + 82.30$$

This illustrates the one-way ANOVA identity.

Below the ANOVA table is another table that gives the sample sizes, sample means, and sample standard deviations of the four samples. Below that table we find the item `Pooled Stdev = 2.268`. This is the pooled estimate of the common population standard deviation, σ, of the four populations.

Finally, the lower right side of the output gives individual 95% confidence intervals for the means, μ_1, μ_2, μ_3, and μ_4, of the four populations under consideration. The formula used to obtain those confidence intervals is presented as exercises in *IS/6e* and *ES/5e*.

PROBLEM 16.2

A chain of convenience stores wanted to test three different advertising policies:

Policy 1: No advertising.
Policy 2: Advertise in neighborhoods with circulars.
Policy 3: Use circulars and advertise in newspapers.

Eighteen stores were randomly selected and divided randomly into three groups of six stores. Each group used one of the three policies. Following the implementation of the policies, sales figures were obtained for each of the stores during a 1-month period. The figures are displayed in the following table in thousands of dollars.

Policy 1	Policy 2	Policy 3
22	21	29
20	25	24
26	25	31
21	20	32
24	22	26
22	26	27

Do the data provide sufficient evidence to conclude that mean sales for the three advertising policies are not all the same? Use $\alpha = 0.05$. Apply Minitab to perform the appropriate hypothesis test. Use the output you obtain to find the three sums of squares, $SSTR$, SSE, and SST, for the sales data. ◇

16.3 Multiple comparisons

I-16.4

When the null hypothesis is rejected by a one-way analysis of variance, the conclusion is that the means are not all equal. The next thing we would want to know is which means are different. More generally, we might want to discover various relationships between the means. Methods for dealing with such problems are called *multiple comparisons.*

The *Tukey multiple-comparison method* is one approach for dealing with multiple comparisons. Similar to the pooled-t confidence interval method for comparing two means, the Tukey method constructs confidence intervals for all possible pairs of means, with the confidence applying simultaneously to all comparisons. This confidence is referred to as the *family confidence level* in *IS/6e* .

Minitab has a subcommand for finding Tukey confidence intervals. However, this subcommand requires that the data are in stacked form. That is, rather than have values for each population in a separate column,as in Example 6.2, all of the values are placed in a single column and matched with a region to distinguish different populations in a second column. Minitab has a way of stacking unstacked data as well as unstacking stacked data in the Manip menu command.

Once the data are in stacked form, you have only to choose Oneway... instead of Oneway (Unstacked)... as in the last example. The corresponding session command is Oneway rather than AOVOneway. We illustrate with the following example.

EXAMPLE 16.3 *Illustrates the ONEWAY command with the TUKEY subcommand*

Use Minitab to perform the Tukey multiple-comparison method to the energy-consumption data of Example 16.2. Use a confidence level of 95%.

SOLUTION First, name two empty columns CONSUME and REGION. With the data already in separate columns, we may stack the data into a single column, matched with a region, using some special options in Minitab. Proceed as follows.

Menu commands:

1 Choose Manip ➤ Stack ➤ Stack columns...
2 Select Stack the following columns: text box and enter NRTHEAST MIDWEST SOUTH WEST (in that order)
3 In the Store the stacked data in: field, check Column of current worksheet and enter CONSUME in the text box
4 Select the Store subscripts in: text box and type REGION
5 Check the Use variable name in subscript column check box
6 Choose Stat ➤ ANOVA ➤ Oneway...
7 Enter CONSUME in the Response: text box
8 Press Tab and enter REGION in the Factor: text box

9 Click the Comparisons... option button and check the Tukey's, family error rate: check box accepting the default value of 5

10 Select OK twice

The corresponding session commands are as follows.

Session commands:

- Type Stack 'NRTHEAST' 'MIDWEST' 'SOUTH' 'WEST' 'CONSUME'; at the MTB > prompt
- Type Subscripts 'REGION'.
- Type ONEWAY 'CONSUME' 'REGION'; at the MTB > prompt and press Enter
- Type TUKEY 5. at the SUBC> prompt and press Enter

That part of the output pertinent to the present discussion is displayed in Printout 16.4.

PRINTOUT 16.4 Minitab output for the Tukey subcommand to the ONEWAY command

```
Tukey's pairwise comparisons

    Family error rate = 0.0500
Individual error rate = 0.0113

Critical value = 4.05

Intervals for (column level mean) - (row level mean)

            MIDWEST     NRTHEAST    SOUTH

NRTHEAST     -2.433
              5.433

SOUTH         0.307       -1.357
              8.693        7.357

WEST          1.367       -0.308      -3.557
              9.233        7.908       5.157
```

The first and second items give the family error rate (default 5% in the example) and the individual error rate that applies to each level. Recall that the confidence level is 1 minus the error rate. The next item in the printout, Critical value = 4.05 is the value of q_a used by the formula at Step 3 in Procedure 16.2 in *IS/6e* .

The final item in the printout provides the endpoints of the confidence intervals for the differences between means corresponding to the levels represented by the row and column headings. Thus, the confidence interval for the difference between population means for the Northeast and South runs from −1.357 to 7.357. Compare this with the table obtained by hand, Table 16.9 in *IS/6e* .

A difference between population means is declared significant if its confidence interval *fails* to include 0. Thus, we may say that there is a significant difference between the Midwest and South as well as between the Midwest and West but no other pairs exhibit a significant difference. ■

PROBLEM 16.3 For the data of Problem 16.2, use Minitab to perform a multiple comparison using a family confidence level of 0.95. ◇

16.4 The Kruskal-Wallis test

I-16.5

The Kruskal-Wallis test is a nonparametric alternative to the one-way ANOVA procedure and is discussed in Section 16.5 in *IS/6e* . Minitab has a command called KRUSKAL-WALLIS, or simply KRUSKAL, that performs the test as outlined in Procedure 16.3 in *IS/6e* .

EXAMPLE 16.4 Illustrates the KRUSKAL command

The U.S. Federal Highway Administration conducts annual surveys on motor vehicle travel by type of vehicle and publishes its findings in *Highway Statistics Summary*. Independent random samples of cars, buses, and trucks provided the data on number of miles driven last year, in thousands, shown in Table 16.2.

TABLE 16.2 Number of miles drive (1000s) last year fro independent samples of cars, buses, and trucks

Cars	Buses	Trucks
19.9	1.8	24.6
15.3	7.2	21.2
2.2	7.2	21.2
6.8	6.5	23.6
34.2	13.3	23.0
8.3	25.4	15.3
12.0		57.1
7.0		14.5
9.5		26.0
1.1		

Use Minitab to perform the hypothesis test the hypothesis that the mean numbers of miles driven are equal at the 5% level of significance.

SOLUTION First, store the mileages from Table 16.2 into a column named `MILES` in stacked form. Accordingly, in a column named `VEHICLE` enter the corresponding names Cars for each data value in the first column of Table 16.2, Buses for data in the second column of Table 16.2, and Trucks for each value in the third column of Table 16.2. Then proceed in the following manner.

Menu commands:

1 Choose Stat ➤ Nonparametrics ➤ Kruskal-Wallis...
2 Enter MILES in the Response: text box
3 Select the Factor: text box and enter VEHICLE
4 Select [OK]

Session commands:

- Type KRUSKAL 'MILES' 'VEHICLE' at the MTB > prompt and press [Enter↵]

The output is shown in Printout 16.5.

PRINTOUT 16.5
Minitab output of the KRUSKAL command

```
Kruskal-Wallis Test on MILES

VEHICLE     N    Median    Ave Rank        Z
Buses       6     7.200         9.0    -1.53
Cars       10     8.900         9.9    -1.75
Trucks      9    23.600        19.2     3.14
Overall    25                  13.0

H = 9.92  DF = 2  P = 0.007
```

The first column of the output, headed VEHICLE, displays the three types of vehicles. The next three columns give the sizes, medians, and mean ranks for the three samples.

The fifth column, headed Z VALUE, displays the population z-scores for the sample mean ranks. This provides a standardized measure of how much each sample mean rank differs from its expected mean rank if the null hypothesis is true (all means are equal). This means that, if the null hypothesis is true, each z-value should be near 0. To the extent that they are not, we would tend to discredit the null hypothesis. This will all be reflected in the value of the test statistic.

In the last line of Printout 16.5, we find the value of the Kruskal-Wallis test statistic, H, the degrees of freedom, and the P-value for the hypothesis test. Since the P-value is 0.007, there is sufficient evidence to conclude that the means are different at the 5% level of significance. ■

PROBLEM 16.4 Use Minitab to perform the Kruskal-Wallis test for the data of Problem 16.2. Use $\alpha = 0.05$. ◇

APPENDIX A

Answers to Problems

In this appendix you will find the answers to all of the problems given in this manual. The answers are grouped by Lessons according to the problem numbers given in the lesson. This should make it easy for you to find the problem you are interested in. Most of the time the answers will involve Minitab printouts. Occasionally you may encounter slight differences in the printout on your screen. These differences come about from the version of Minitab you may be using and the one we used to find the answers.

Lesson 1

PROBLEM 1.1

We chose C50. Minitab output:

```
MTB > SET C50
DATA> 48 41 57 83 41 55 59
DATA> 61 38 48 79 75 77  7
DATA> 54 23 47 56 79 68 61
DATA> 64 45 53 82 68 38 70
DATA> 10 60 83 76 21 65 47
DATA> END
MTB >NAME C50 'DIABETES'
```

PROBLEM 1.2

We chose C51, C52. Minitab output:

```
MTB > READ C51 C52
DATA> 10644 2995
DATA> 11707 623
DATA> 9360 416
DATA> 6883 286
DATA> 1905 497
DATA> 3284 26
MTB > END
      6 ROWS READ
```

PROBLEM 1.3

Minitab output:

```
MTB > PRINT 'DIABETES'

DIABETES
   48  41  57  83  41  55  59  61  38  48  79  75  77
    7  54  23  47  56  79  68  61  64  45  53  82  68
   38  70  10  60  83  76  21  65  47
```

PROBLEM 1.4

Minitab output:

```
MTB > SAVE 'DIABETICS'
Worksheet saved into file: DIABETICS.MTW
MTB > RESTART
MTB > RETRIEVE 'DIABETICS'
 Retrieving worksheet from file: DIABETICS.MTW
# Worksheet was saved on Thu Mar 29 2001
```

Continued on the following page.

```
Worksheet retrieved from file: DIABETICS.MTW
MTB > PRINT 'DIABETES'

DIABETES
   48  41  57  83  41  55  59  61  38  48  79  75  77
    7  54  23  47  56  79  68  61  64  45  53  82  68
   38  70  10  60  83  76  21  65  47
```

PROBLEM 1.5 Minitab output:

```
MTB > RESTART
MTB > READ 'CONT' C53 C54
Entering data from file: CONT.DAT
      6 rows read.
MTB > PRINT C53 C54

 Row      C53     C54

   1    10644    2995
   2    11707     623
   3     9360     416
   4     6883     286
   5     1905     497
   6     3284      26
```

PROBLEM 1.6 Answers will vary.

PROBLEM 1.7 Answers will vary.

Lesson 2

PROBLEM 2.1 Minitab output:

```
 Grouped-data table for DAYS      N=  40

Row  LowerCut  UpperCut   Freq  RelFreq  Midpoint

  1        30        40      3    0.075        35
  2        40        50      1    0.025        45
  3        50        60      8    0.200        55
  4        60        70     10    0.250        65
  5        70        80      7    0.175        75
  6        80        90      7    0.175        85
  7        90       100      4    0.100        95
```

PROBLEM 2.2

Minitab output:

PERSONS	Count	Percent
1	7	17.50
2	13	32.50
3	9	22.50
4	5	12.50
5	4	10.00
6	1	2.50
7	1	2.50
N=	40	

PROBLEM 2.3

Minitab output:

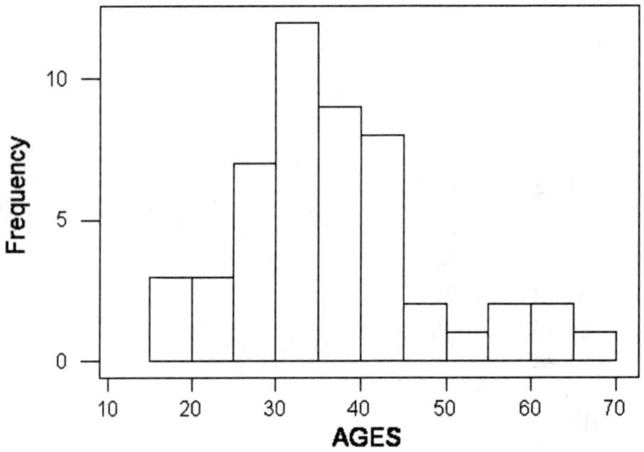

Minitab character graph:

```
Histogram of AGES   N = 50

Midpoint   Count
   17.00       3  ***
   22.00       3  ***
   27.00       7  *******
   32.00      12  ************
   37.00       9  *********
   42.00       8  ********
   47.00       2  **
   52.00       1  *
   57.00       2  **
   62.00       2  **
   67.00       1  *
```

PROBLEM 2.4 Minitab output:

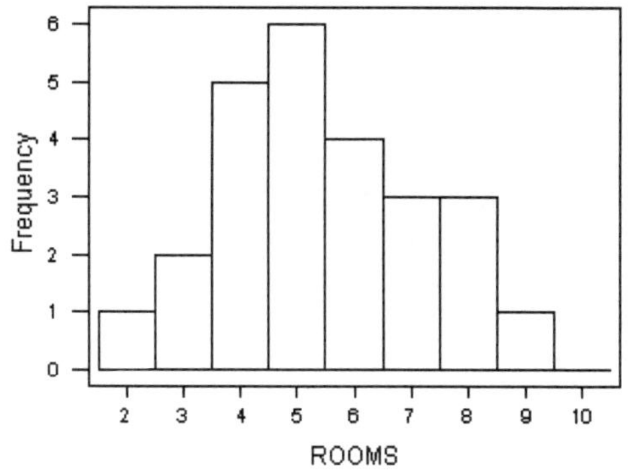

Minitab character graph:

```
Histogram of ROOMS   N = 25

Midpoint   Count
    2.00       1  *
    3.00       2  **
    4.00       5  *****
    5.00       6  ******
    6.00       4  ****
    7.00       3  ***
    8.00       3  ***
    9.00       1  *
```

PROBLEM 2.5 Minitab output:

```
MTB > DOTPLOT 'STRENGTH';
SUBC> INCREMENT=1;
SUBC> START=9, 11.
                   ..
          .  . .::  :.
        +---------+---------+-STRENGTH
      9.00     10.00     11.00
```

PROBLEM 2.6 a) Minitab output:Using an increment of 10:

```
Stem-and-leaf of AGES      N  = 50
Leaf Unit = 1.0

    3    1 699
   13    2 1246678899
  (21)   3 011223333444557778889
   16    4 0012233359
    6    5 178
    3    6 025
```

b) Minitab output:Using an increment of 5:

```
Stem-and-leaf of AGES      N  = 50
Leaf Unit = 1.0

    3    1 699
    6    2 124
   13    2 6678899
   25    3 011223333444
   25    3 557778889
   16    4 00122333
    8    4 59
    6    5 1
    5    5 78
    3    6 02
    1    6 5
```

Lesson 3

PROBLEM 3.1 Minitab output:

```
Mean of LT2    =      63.500
Mean of GE2    =      74.778
```

PROBLEM 3.2 Minitab output:

```
Median of LT2    =      62.500
Median of GE2    =      75.000
```

PROBLEM 3.3 Enter the data in a column and use HUD for the name of the column. Minitab output:

```
HUD  Count
  2      1
  3      2
  4      5
  5      6
  6      4
  7      3
  8      3
  9      1
 N=     25
```

The mode of the number of rooms is 5.

PROBLEM 3.4 Minitab output:

```
Sum of PATIENTS=      229.30
```

Hence the total number of patients admitted to U.S. hospitals between 1977 and 1982 was 229.3 million.

PROBLEM 3.5 Assume that there is still a column named XBAR. Minitab output:

```
MTB > LET 'XBAR'=SUM('FBI')/N('FBI')
MTB > PRINT 'XBAR'
XBAR
   78.5

  Mean of FBI      =       78.500
```

PROBLEM 3.6 With the data in FBI, Minitab output:

```
  Range of FBI      =       141.00
```

PROBLEM 3.7 Minitab output:

```
  Standard deviation of FBI      =       39.212
```

PROBLEM 3.8 Minitab output:

Variable	Minimum	Maximum	Q1	Q3
ENERGY	949.0	1532.0	1103.8	1389.5

Hence, $IQR = 285.8$

PROBLEM 3.9 Minitab Describe output:

Variable	N	Mean	Median	TrMean	StDev	SE Mean
ENERGY	36	1244.2	1231.0	1245.4	165.5	27.6

Variable	Minimum	Maximum	Q1	Q3
ENERGY	949.0	1532.0	1103.8	1389.5

Minitab graphics output:

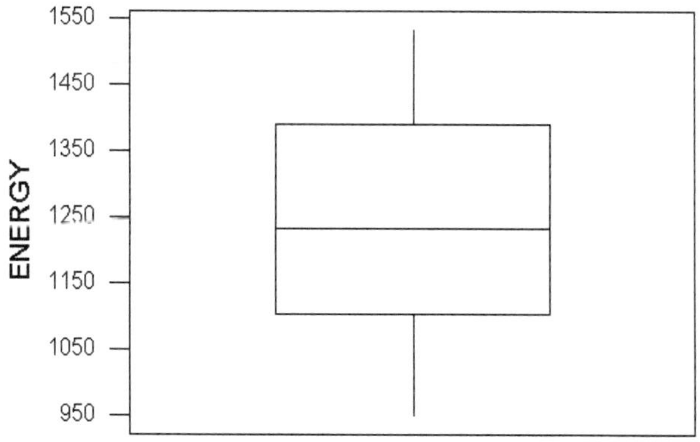

The box extends from 1103.7 to 1389.5 with a horizontal line at 1231. IQR=285.8 so the bottom whisker extends to the minimum value 949 and the top whisker extends to the maximum value 1532. There are no potential outliers.

PROBLEM 3.10 Minitab output:

```
Mean of HTS = 74.057
```

Hence, the population mean height is $\mu = 75.000$ inches.

PROBLEM 3.11 Minitab output:

```
MTB > LET K26=SQRT((N('HTS')-1)/N('HTS'))*STDEV('HTS')
MTB > PRINT K26
K26     2.92945
```

Consequently, the population standard deviation of the heights of the 11 players is 2.29624 inches.

Lesson 4

PROBLEM 4.1 a) Minitab output:

```
MTB > let K2=('f'(1)+'f'(2))/sum('f')
MTB > PRINT K2
K2      0.109597
```

b) Minitab output:

```
MTB > let K2=('f'(2)+'f'(3)+'f'(4))/sum('f')
MTB > PRINT K2
K2     0.297589
```

c) Minitab output:

```
MTB > let K2=('f'(1)+'f'(7))/sum('f')
MTB > PRINT K2
K2     0.312840
```

PROBLEM 4.2 Minitab output:

```
Rows: YEARS      Columns: PARTY
          Democrat Republic       All

Under 2       59        25        84
2-9          134        93       227
10-19         44        41        85
20-29         20         8        28
30&over       10         0        10
All          267       167       434

 Cell Contents --             Count
```

PROBLEM 4.3

Minitab output:

Rows: YEARS Columns: PARTY

	Democrat	Republic	All
Under 2	13.59	5.76	19.35
2-9	30.88	21.43	52.30
10-19	10.14	9.45	19.59
20-29	4.61	1.84	6.45
30&over	2.30	--	2.30
All	61.52	38.48	100.00

Cell Contents -- % of Tbl

The probability of obtaining a congressman age 20–29 is 0.0645.

PROBLEM 4.4

Minitab output:

Rows: YEARS Columns: PARTY

	Democrat	Republic	All
Under 2	22.10	14.97	19.35
2-9	50.19	55.69	52.30
10-19	16.48	24.55	19.59
20-29	7.49	4.79	6.45
30&over	3.75	--	2.30
All	100.00	100.00	100.00

Cell Contents -- % of Col

Rows: YEARS Columns: PARTY

	Democrat	Republic	All
Under 2	70.24	29.76	100.00
2-9	59.03	40.97	100.00
10-19	51.76	48.24	100.00
20-29	71.43	28.57	100.00
30&over	100.00	--	100.00
All	61.52	38.48	100.00

Cell Contents -- % of Row

The probability of a Republican among those in office for under 2 years is 0.1479; the probability of a congressman being in office under 2 years among Democrats is 0.7024.

Lesson 5

PROBLEM 5.1

Enter the data into z and fz as in Example 5.1. Minitab output:

```
ROW   z        P(Z=z)
  1   1   0.232057
  2   2   0.316986
  3   3   0.174641
  4   4   0.154306
  5   5   0.072967
  6   6   0.029904
  7   7   0.019139
```

PROBLEM 5.2

Answers will vary.

PROBLEM 5.3

Minitab output:

```
MTB > LET K21=SUM('z'*'P(Z=z)')
MTB > PRINT K21
K21      2.68541
```

Thus the mean of the random variable Z is $\mu_z = 2.68541$ persons.

PROBLEM 5.4

Minitab output:

```
MTB > LET K22=SQRT(SUM(('z'-K21)**2*'P(Z=z)'))
MTB > PRINT K22
K22      1.47234
```

Thus the standard deviation of the random variable Z is $\sigma = 1.4723$ persons.

PROBLEM 5.5

Minitab output of macro %MUSIG 'z' 'P(Z=z)':

```
mu       2.68541
sigma    1.47234
```

Both the mean and standard deviation agree with those obtained previously.

PROBLEM 5.6

a) Minitab output:

```
Binomial with n = 3 and p = 0.250000
           x      P( X = x)
        2.00         0.1406
```

Therefore, the probability is 0.1406 that exactly two of the three calls last longer than 3.8 minutes.

b) Minitab output:

```
Binomial with n = 3 and p = 0.250000
           x     P( X <= x)
        2.00         0.9844
```

Thus, the probability is 0.9844 that at most two of the three calls last longer than 3.8 minutes.

c) Minitab output:

```
Binomial with n = 3 and p = 0.250000
         x      P( X <= x)
      0.00          0.4219
 MTB > LET K1=1-0.4219
 MTB > PRINT K1
 K1        0.578100
```

d) Minitab output:

```
 Row      x          Px

   1      0   0.421875
   2      1   0.421875
   3      2   0.140625
   4      3   0.015625
```

e) Minitab graph output:

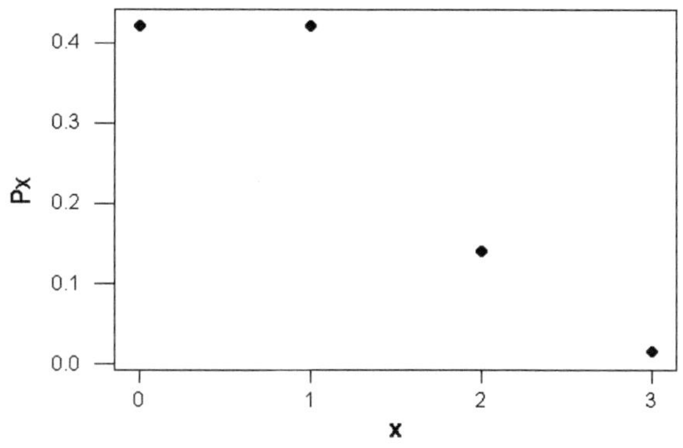

The distribution is skewed to the right.

PROBLEM 5.7 Enter the distribution in x and Px. Minitab output:

```
MTB > %MUSIG 'x' 'Px'
mu         0.750000
sigma      0.750000
```

PROBLEM 5.8 a) Minitab output:

```
Poisson with mu = 1.00000
         x       P( X = x)
      2.00          0.1839
```

b) Minitab output:

```
Poisson with mu = 1.00000

          x      P( X <= x)
       2.00          0.9197
```

c) Minitab output:

```
Poisson with mu = 1.00000

          x       P( X = x)
          0          0.3679
          1          0.3679
          2          0.1839
          3          0.0613
          4          0.0153
          5          0.0031
          6          0.0005
          7          0.0001
          8          0.0000
```

d Minitab graph output:

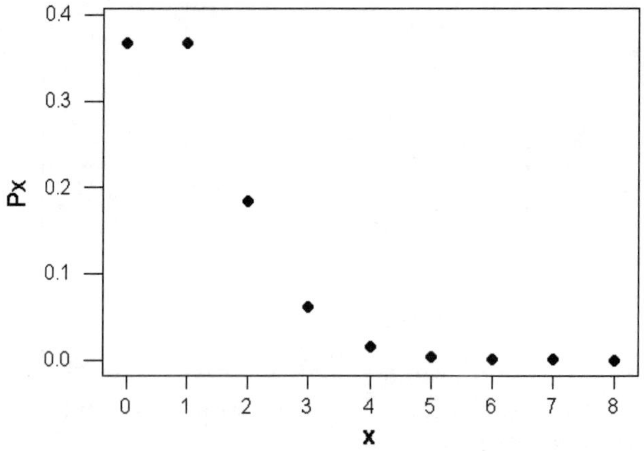

The distribution is skewed to the right.

PROBLEM 5.9 Minitab output:

```
MTB > %musig 'x' 'Px'

mu        0.999990
sigma     0.999964
```

The answers agree except for rounding partly because the entire distribution is not recorded in x and Px.

PROBLEM 5.10

a) Minitab output:

```
Poisson with mu = 2.25000

          x      P( X = x )
       2.00          0.2668
Binomial with n = 500 and p = 0.00450000

          x      P( X = x )
       2.00          0.2673
```

b) Minitab output:

```
Poisson with mu = 2.25000

          x     P( X <= x )
       4.00          0.9220
Binomial with n = 500 and p = 0.00450000

          x     P( X <= x )
       4.00          0.9224
```

Lesson 6

PROBLEM 6.1

Answers will vary.

PROBLEM 6.2

a) Minitab output:

```
Normal with mean = 0 and standard deviation = 1.00000
          x      P( X <= x)
    -1.2800          0.1003
```

The area under the standard normal curve to the left of $z = -1.28$ equals 0.1003.

b) Minitab output:

```
MTB > SET C23
DATA> 1.53 2.47
DATA> END
MTB > CDF C23 C24
MTB > LET K21=C24(2)-C24(1)
MTB > PRINT K21
K21      0.0562527
```

The area between $z = 1.53$ and $z = 2.47$ equals 0.0563.

c) Minitab output:

```
MTB > CDF 1.64 K22
MTB > LET K23=1-K22
MTB > PRINT K23
K23       0.0505026
```

The area under the standard normal curve to the right of $z = 1.64$ equals 0.0505026.

d) Minitab output:

```
MTB > SET C25
DATA> -1.85 0.29
DATA> END
MTB > CDF C25 C26
MTB > LET K24=C26(1)+1-C26(2)
MTB > PRINT K24
K24       0.418065
```

The area under the standard normal curve that lies either to the left of $z = -1.85$ or to the right of $z = 0.29$ equals 0.418065.

PROBLEM 6.3

a) Minitab output:

```
Normal with mean = 0 and standard deviation = 1.00000
 P( X <= x)           x
    0.7500        0.6745
```

Hence, $z_{0.25} = 0.6745$; that is, an area of 0.75 is to the left of $z = 0.6745$ under the standard normal curve.

b) Minitab output:

```
Normal with mean = 0 and standard deviation = 1.00000
 P( X <= x)           x
    0.0500       -1.6449
```

Thus, $z_{0.95} = -1.6449$; that is, an area of 0.05 is to the left of $z = -1.6449$ under the standard normal curve.

c) Minitab output:

```
Normal with mean = 0 and standard deviation = 1.00000
 P( X <= x)           x
    0.9500        1.6449
```

Thus, $z_{0.05} = 1.6449$; that is, an area of 0.05 is to the right of $z = 1.6449$ under the standard normal curve. Combining this with b), there is a middle area of 90% between $z = -1.6449$ and $z = 1.6449$. Note that $z_{0.95} = -z_{0.05}$.

d) Minitab output:

```
Normal with mean = 0 and standard deviation = 1.00000
 P( X <= x)           x
    0.9400        1.5548
```

So, $z_{0.06}$=1.5548 and, as in c), $z_{0.94} = -1.5548$.

PROBLEM 6.4

a) Minitab output:

```
Normal with mean = 15.0000 and standard deviation = 2.00000
          x      P( X <= x)
   12.0000          0.0668
```

Thus, the area under the normal curve with parameters $\mu = 15$ and $\sigma = 2$ that lies to the left of 12 equals 0.0668.

b) Minitab output:

```
MTB > SET C31
DATA> 14 18
DATA> END
MTB > CDF C31 C32;
SUBC> NORMAL 15 2.
MTB > LET K25=C32(2)-C32(1)
MTB > PRINT K25
K25       0.624655
```

Consequently, the area under the normal curve with parameters $\mu = 15$ and $\sigma = 2$ that lies between 14 and 18 equals 0.624655.

c) Minitab output:

```
MTB > CDF 16.5 K26;
SUBC> NORMAL 15 2.
MTB > LET K27=1-K26
MTB > PRINT K27
K27       0.226627
```

Hence, the area under the normal curve with parameters $\mu = 15$ and $\sigma = 2$ that lies to the right of 16.5 equals 0.226627.

PROBLEM 6.5

a) Minitab output:

```
Normal with mean = 64.6000 and standard deviation = 2.40000
 P( X <= x)            x
    0.9000       67.6757
```

Thus, $x_{0.10} = 67.6757$; that is, there is an area of 0.90 to the left of $x = 67.6757$ under a normal curve with parameters $\mu = 64.6$ and $\sigma = 2.4$.

b) Minitab output:

```
Normal with mean = 64.6000 and standard deviation = 2.40000
 P( X <= x)            x
    0.1000       61.5243
```

Hence, $x_{0.90} = 61.5243$; that is, there is an area of 0.90 to the right of the point $x = 61.5243$ under a normal curve with parameters $\mu = 64.6$ and $\sigma = 2.4$.

PROBLEM 6.6 Minitab graph output:

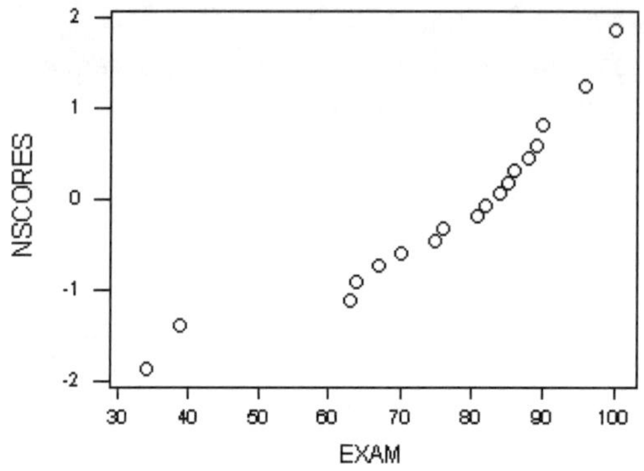

The plot is reasonably linear.

PROBLEM 6.7 Minitab output:

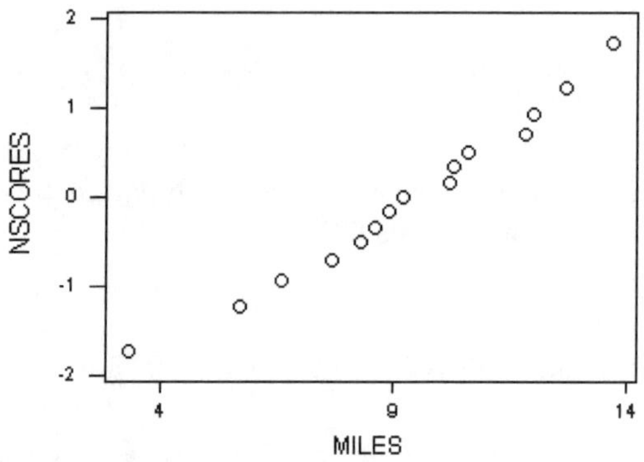

The plot is reasonably linear; there is a possibility that the smallest observation may be an outlier. If that value is omitted from the data set, then you will find linearity is slightly degraded, not improved by removing the minimum value.

Lesson 7

PROBLEM 7.1 Answers will vary.

PROBLEM 7.2 Answers will vary.

PROBLEM 7.3 Answers will vary.

Lesson 8

PROBLEM 8.1

Minitab output:

```
The assumed sigma = 12.8

Variable          N       Mean     StDev   SE Mean          90.0% CI
WEIGHT           40     136.88     12.77      2.02  (  133.55, 140.20)
```

From the last item of the output, we see that a 90% confidence interval for μ is from 133.55 to 140.20. We can be 90% confident that the mean weight, μ, of all U.S. women 5 feet 4 inches tall and in the age group 18–24 is somewhere between 133.55 and 140.20 pounds.

PROBLEM 8.2

a) Minitab output:

```
Student's t distribution with 8 d.f.
 P( X <= x)            x
    0.9000        1.3968
```

Thus, 1.3968 is the require t-value; that is, $t_{0.10} = 1.3968$.

b) Minitab output:

```
Student's t distribution with 8 d.f.
 P( X <= x)            x
    0.9950        3.3554
```

From the output, $t_{0.005} = 3.3554$.

c) Minitab output:

```
Student's t distribution with 8 d.f.
 P( X <= x)            x
    0.0100       -2.8965
```

Hence, -2.8965 is the t-value having area 0.01 to its left; that is, $t_{0.99} = -2.8965$.

PROBLEM 8.3

Minitab output:

```
Variable      N      Mean     StDev  SE Mean        95.0 % CI
SUBRATES     12    122.75      8.44     2.44  (  117.39,   128.11)
```

From the last item of the output, we see that a 95% confidence interval for μ is from 117.39 to 128.11. We can be 95% confident that the mean weight, μ of this year's mean annual subscription rate is somewhere between \$117.39 and \$128.11.

Lesson 9

PROBLEM 9.1

Minitab output:

```
Test of mu = 13 vs mu not = 13
The assumed sigma = 11.6

Variable          N      Mean     StDev   SE Mean
TRAVTIME         35     14.03     13.32      1.96

Variable             95.0% CI            Z       P
TRAVTIME       (   10.19,    17.87)    0.52  0.600
```

The output shows that the P-value is 0.60; 13 is included in the confidence interval; the nonrejection region is $-1.96, +1.96$ and this includes the Z-value of 0.52. On all accounts, we do not reject H_0. The data do not provide sufficient evidence to conclude that this year's mean travel time to work for South Dakota residents have changed from the 1980 mean of 13 minutes.

PROBLEM 9.2

Minitab output:

```
Test of mu = 428 vs mu > 428

Variable          N      Mean     StDev   SE Mean
SAT              25     437.4      86.7      17.3

Variable       90.0% Lower Bound        T       P
SAT                        414.6     0.54   0.296
```

The output shows that the P-value is 0.30; a lower one-sided confidence interval is $(414.6, \infty)$ which includes 428 and, with 24 df, the rejection region is defined by $(1.3178, \infty)$ and does not include the value of the test statistic, 0.54. By all accounts, do not reject H_0. The data do not provide sufficient evidence to conclude that last year's mean for verbal SAT scores has increased over the 1983 mean of 428 points.

PROBLEM 9.3

Minitab output:

```
Stem-and-leaf of TVPRICE     N  = 8
Leaf Unit = 10
    2     5 00
    2     5
    3     5 5
    4     5 6
    4     5 8
    3     6 1
    2     6
    2     6 44
```

There being no outstanding evidence of lack of symmetry, we proceed:

```
Test of median = 599.0 versus median  <  599.0

                     N for   Wilcoxon            Estimated
             N   Test  Statistic        P      Median
TVPRICE      8      8       13.0    0.264       575.3
----------------------------------------------------------

              Estimated    Achieved
             N    Median  Confidence  Confidence Interval
TVPRICE      8     575.3        89.3  (   537.0,    614.0)
```

With such a high P-value, the data do not support a change in price. A 95% upper one-sided confidence interval is $(-\infty, 614.0)$ and 575.3 does fall in that interval.

Lesson 10

PROBLEM 10.1 Minitab output:

```
Two-sample T for NEW vs PRESENT

          N      Mean     StDev   SE Mean
NEW      10    42.130     0.685      0.22
PRESENT  10    43.230     0.750      0.24

Difference = mu NEW - mu PRESENT
Estimate for difference:  -1.100
95% upper bound for difference: -0.543
T-Test of difference = 0 (vs <): T-Value = -3.43  P-Value = 0.002  DF = 18
Both use Pooled StDev = 0.718
```

The output shows that the P-value is 0.002; the 95% upper one-sided confidence interval is $(-\infty, -0.543)$; with `df=18`, the single critical value is $-t_{.05} = -1.7341$. On all accounts, we reject H_0. Evidently, the new machine packs faster, on the average.

PROBLEM 10.2 Minitab output:

```
Two-sample T for SURE vs MIRROR

          N      Mean     StDev   SE Mean
SURE     15     89.93      1.94      0.50
MIRROR   15     92.27      1.03      0.27

Difference = mu SURE - mu MIRROR
Estimate for difference:  -2.333
99% upper bound for difference: -0.902
T-Test of difference = 0 (vs <): T-Value = -4.10  P-Value = 0.000  DF = 2121
```

We see from the output that the P-value is 0.000. Since this is less than the designated significance level of $\alpha = 0.01$, we reject H_0. Also, a 99% upper one-sided confidence interval for the difference is $(-\infty, -0.902)$ and that excludes 0. Finally, the single critical value is $-t_{.01} = -2.5176$ and the `T-Value` of -4.10 is less than that. The data indicate that Mirror-Sheen has a longer effectiveness time, on the average, than Sureglow.

PROBLEM 10.3

Minitab output:

```
MEN          N =   8      Median =          408.5
WOMEN        N =  10      Median =          280.5
Point estimate for ETA1-ETA2 is            117.5
91.7 Percent CI for ETA1-ETA2 is (-0.0,243.0)
W = 95.5
Test of ETA1 = ETA2  vs  ETA1 > ETA2 is significant at 0.0457
The test is significant at 0.0456 (adjusted for ties)
```

At the 5% level of significance (barely), you should reject H_0; the data suggest that male full-time wages are greater than those for females. The 91.7% lower one-sided confidence interval would, by this computation be $(-0.0, \infty)$ which barely includes 0. If you rework the problem using a confidence level of 89%, you would discover that the 90% lower one-sided confidence interval is $(23.1, \infty)$ and this definitely excludes 0 confirming rejection of the null hypothesis.

PROBLEM 10.4

Minitab output for TTEST:

```
Paired T for CURRENTV - NEWV
                 N       Mean      StDev    SE Mean
CURRENTV        10      37.40       3.72       1.18
NEWV            10      38.30       4.32       1.37
Difference      10     -0.900      2.514      0.795

95% upper bound for mean difference: 0.558
T-Test of mean difference = 0 (vs < 0): T-Value = -1.13  P-Value = 0.143
```

From the output, we see that the P-value is 0.143; the 95% upper one-sided confidence interval is $(-\infty, 0.558)$; the single critical value, with `df=9` is $-t_{.05} = -1.8331$. By all accounts, we do not reject H_0. The data do not provide sufficient evidence to conclude that the new variety of wheat gives a better yield, on the average, than the current variety.

PROBLEM 10.5

Minitab output:

```
Test of median = 0.000000 versus median  <  0.000000

                 N for   Wilcoxon              Estimated
            N     Test  Statistic          P      Median
DIFF       10        9       11.0      0.096      -1.500
```

At the 5% level of significance, the evidence does not support a difference in the yields.

Lesson 11

PROBLEM 11.1

a) Minitab output:

```
MTB > INVCDF .10;
SUBC> CHIS 14.
P( X <= x)          x
   0.1000       7.7895
```

b) Minitab output:

```
MTB > INVCDF .9;
SUBC> CHIS 14.
  Chi-Square with 14 DF
    P( X <= x)          x
       0.9000       21.0641
```

PROBLEM 11.2

a) Minitab output:

```
MTB > CDF 13 K1;
SUBC> CHIS 14.
MTB > LET K2=1-K1
MTB > PRINT K2
K2    0.526524
```

b) Minitab output:

```
MTB > CDF 7 K3;
SUBC> CHIS 14.
MTB > LET K4=K1-K3
MTB > PRINT K4
K4    0.408188
```

PROBLEM 11.3

Minitab output of %1stdev 'VERBAL':

```
Test of sigma =    100 vs sigma not =    100
Row  Variable    n    StDev  Chi-Sq       P
  1    VERBAL   25  110.982  29.561  0.399
```

From the output, we cannot conclude that the standard deviation of this year's verbal scores is different from the 1941 standard deviation of 100.

PROBLEM 11.4

Minitab output of %1stdev 'VERBAL':

```
  Row  Variable     n    StDev  Level         CI for sigma

    1    VERBAL    25  110.982  95.0%  (86.658, 154.393)
```

From the output, we may say that the standard deviation of this year's verbal scores is somewhere between 87 and 154 with confidence 95%. This interval includes the 1941 standard deviation of 100 by the way.

PROBLEM 11.5

a) Minitab output:

```
MTB > INVCDF .95;
SUBC> F 12 5.
F distribution with 12 DF in numerator and 5 DF in denominator
P( X <= x)           x
     0.9500     4.6777
```

b) Minitab output:

```
MTB > INVCDF .99;
SUBC> F 12 5.
 P( X <= x)           x
    0.9900     9.8883
```

c) Minitab output:

```
MTB > INVCDF .975;
SUBC> F 12 5.
 P( X <= x)           x
    0.9750     6.5245
```

d) Minitab output:

```
MTB > CDF 1.5 K1;
SUBC> F 12 5.
MTB > LET K2=1-K1
MTB > PRINT K2
K2     0.343970
```

PROBLEM 11.6

Minitab output:

```
F-Test of sigma1 = sigma2 (vs <)


Row  Variable    n  StDev      F      P

  1         S   12  0.587   0.19  0.005
  2         U   12  1.346
```

From the output, conclude that there is evidence that rainfall is less variable in regions that were seeded.

PROBLEM 11.7

Minitab output:

```
Row  Variable    n  StDev  Level    CI for ratio

  1         S   12  0.587  90.0%  (0.260, 0.732)
  2         U   12  1.346
```

From the output, an upper one-sided 95% confidence interval for the ratio is given by $(0, 0.732)$ which falls short of including 1. Hence, conclude on this basis that rainfall is less variable in regions that were seeded.

Lesson 12

PROBLEM 12.1

Minitab output:

```
Sample       X      N  Sample p          99.0 % CI          Z-Value  P-Value
 1        1238   1528  0.810209  (0.784369, 0.836049)        24.25    0.000
```

From the output, a 99% confidence interval for p runs from 0.784 to 0.836.

PROBLEM 12.2

Minitab output:

```
                                                              Exact
Sample       X      N  Sample p          95.0 % CI          P-Value
1            9     21  0.428571  (0.218197, 0.659794)         0.664
```

From the output, an exact 95% confidence interval for p runs from 0.218 to 0.660.

PROBLEM 12.3

Minitab output:

```
 Test of p = 0.1 vs p > 0.1

Sample       X      N  Sample p  99.0% Lower Bound  Z-Value  P-Value
1           24    150  0.160000           0.090365     2.45    0.007
```

The null hypothesis should be rejected at the 1% level because the P-value is only 0.007. Yet you would not reject the null hypothesis based on confidence intervals since 0.1 is greater than the lower one-sided 99% confidence bound of 0.09. This is one of those cases where the two procedures are not quite equivalent.

However, if we were to perform the exact test, we would arrive at the Minitab output:

```
Test of p = 0.1 vs p > 0.1

                                                      Exact
Sample       X      N  Sample p  99.0% Lower Bound  P-Value
1           24    150  0.160000           0.096900    0.014
```

With a more exact computation of the P-value, you would not reject by either criteria this time and would conclude that this year's percentage of active licensees is not higher and the two procedures agree.

PROBLEM 12.4 Minitab output:

```
Sample       X        N  Sample p
1          184      300  0.613333
2          148      250  0.592000
Estimate for p(1) - p(2):  0.0213333
95% CI for p(1) - p(2):   (-0.0608138, 0.103480)
Test for p(1) - p(2) = 0 (vs not = 0):  Z = 0.51  P-Value = 0.611
```

With a P-value of 0.611, the evidence is not sufficient to conclude that there was a real difference. We can be 95% confident that the difference, $p_1 - p_2$, is somewhere between -0.061 and 0.103 and this includes 0.

Lesson 13

PROBLEM 13.1 See the solution to Problem 11.1, page A-21.

PROBLEM 13.2 See the solution to Problem 11.2, page A-21.

PROBLEM 13.3 %fittest output:

```
Chi-square goodness-of-fit test

Row       n      k      ChiSq  P-value

   1    250      6    153.645         0
```

From the output, we see that the value of the test statistic is $\chi^2 = 153.645$. From the same printout, we find that the P-value associated with this statistic is 0. Reject H_0. It appears that the primary-heating-fuel distribution for occupied housing units built after 1974 differs from that of all occupied housing units.

PROBLEM 13.4 Minitab output:

```
Rows: SEX      Columns: REPLY

              No        Und        Yes        All

Female     17.50      10.00      17.50      45.00
Male       27.50      15.00      12.50      55.00
All        45.00      25.00      30.00     100.00
  Cell Contents --
                     % of Tbl
```

The chances that a respondent is both Male and answers Yes is 12.5%.

PROBLEM 13.5 Answers will vary.

PROBLEM 13.6 Minitab output:

```
 Expected counts are printed below observed counts

              No       Und       Yes     Total
    1          7         4         7        18
            8.10      4.50      5.40

    2         11         6         5        22
            9.90      5.50      6.60

Total         18        10        12        40

Chi-Sq =  0.149 +  0.056 +  0.474 +
          0.122 +  0.045 +  0.388 = 1.235
DF = 2, P-Value = 0.539
1 cells with expected counts less than 5.0
```

With a P-value of 0.539 the data do not provide sufficient evidence to conclude that the sex of a respondent and their reply on the issue of approval are dependent.

Lesson 14

PROBLEM 14.1 Minitab graph output:

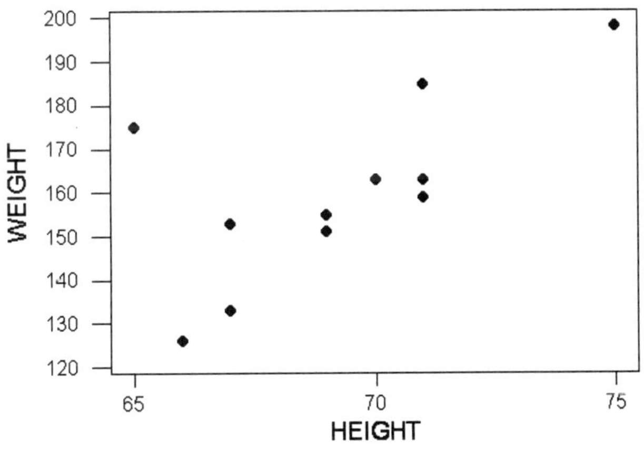

Reasonably linear.

PROBLEM 14.2 Minitab output:

```
The regression equation is
WEIGHT = - 174 + 4.84 HEIGHT

Predictor        Coef    SE Coef          T        P
Constant       -174.5      126.3      -1.38    0.200
HEIGHT          4.836      1.824       2.65    0.026

S = 16.48       R-Sq = 43.9%      R-Sq(adj) = 37.6%

Analysis of Variance

Source            DF        SS         MS        F        P
Regression         1    1909.5     1909.5     7.03    0.026
Residual Error     9    2443.4      271.5
Total             10    4352.9
```

PROBLEM 14.3 From the printout in Problem 14.2, `R-sq = 43.9%`, so $r^2 = 0.439$.

PROBLEM 14.4 From the analysis of variance portion of the printout in Problem 14.2, we have $SSR = 1909.5$, $SSE = 2443.4$ and $SST = 4352.9$. Hence, $r^2 = 1909.5/4352.9 = 0.439$ or 43.9%.

PROBLEM 14.5 Minitab output:

```
Pearson correlation of WEIGHT and HEIGHT = 0.662
P-Value = 0.026
```

Thus, the linear correlation coefficient of the data is 0.662.

Lesson 15

PROBLEM 15.1 a) From the Minitab output for Problem 14.2 we have `s = 16.48`; thus, $s_e = 16.48$.
b) 16.48 lb.

PROBLEM 15.2 Minitab graphics output for residuals:

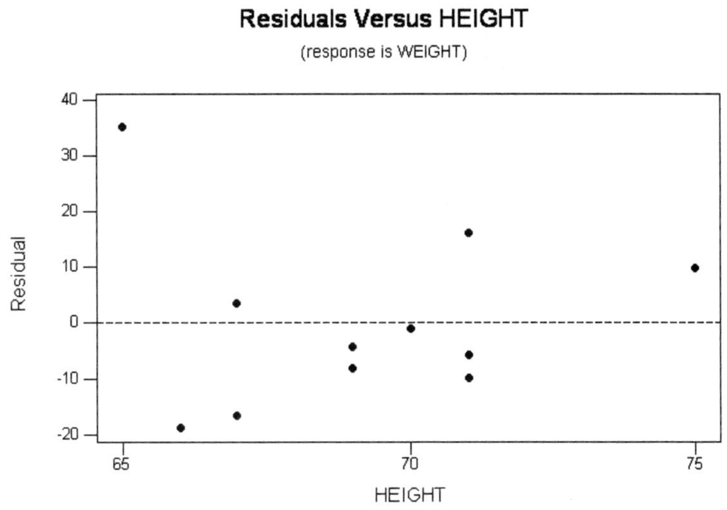

The funnel effect from left to right in the residual plot is a little disturbing, indicating that the variance may be decreasing as height increases. Minitab probability plot:

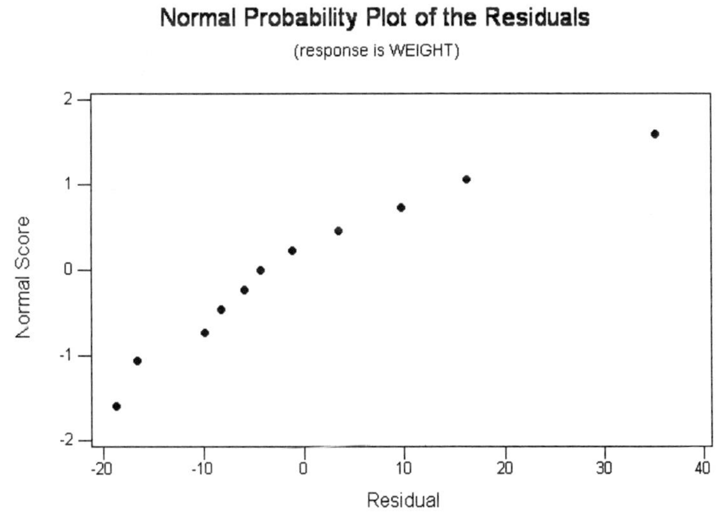

Except for the pattern for the first few observations, the plot is roughly linear.

PROBLEM 15.3

We need to perform the test:

$$H_0\colon \beta_1 = 0 \text{ (height is not useful for predicting weight)}$$
$$H_a\colon \beta_1 \neq 0 \text{ (height is useful for predicting weight)}$$

at the 5% level of significance.
From the fifth line of the Minitab output for Problem 14.2, the line labeled `HEIGHT`, the fourth entry under the column headed `t-ratio`, displays the value of the test statistic, $t = 2.65$. The final entry gives the P-value of 0.026. Since this is less than the designated significance level of $\alpha = 0.05$, we reject H_0. The data provide sufficient evidence to conclude that the slope of the populations regression line is not zero; hence, height is a useful predictor for weight.

PROBLEM 15.4

STEP 1 For a 95% confidence interval, $\alpha = 0.05$. Since $n = 11$, df$=11-2=9$. We need to find $t_{\alpha/2} = t_{0.05/2} = t_{0.025}$ for df $= 9$. We can use Minitab for this task.
Minitab output:

```
Student's t distribution with 9 d.f.

 P( X <= x)          x
    0.9750        2.2622
```

The printout shows that $t_{0.025} = 2.2622$.
STEP 2 The endpoints of the confidence interval are

$$b_1 \pm t_{\alpha/2} \cdot s_e/\sqrt{\Sigma x^2 - (\Sigma x)^2/n}$$

Examine the printout for Problem 14.2. From the line labeled `HEIGHT`, The second entry in that line, which is in the column headed `Coef`, gives b_1. So, $b_1 = 4.836$. The third entry in that line, which is in the column headed `Stdev`, gives the estimated standard deviation of b_1; that is, $s_e/\sqrt{\Sigma x^2 - (\Sigma x)^2/n}$. So,

$$s_e/\sqrt{\Sigma x^2 - (\Sigma x)^2/n} = 1.824$$

Consequently, the endpoints of the confidence interval are

$$4.836 \pm 2.2622 \cdot 1.824$$

or 0.7097 to 8.9623. We can be 95% confident that β_1 is between 0.7097 and 8.9623 pounds.

PROBLEM 15.5

a) Minitab output:

```
 Predicted Values for New Observations
New Obs      Fit      SE Fit          95.0% CI                 95.0% PI
1          164.05       5.19     (  152.31,   175.78)  (  124.97,   203.13)

Values of Predictors for New Observations
New Obs     HEIGHT
1             70.0
```

From the output we see that a 95% confidence interval is from 152.31 to 175.79. We can be 95% confident that the mean weight for the population of males 18–24 years old who are 70 inches tall is somewhere between 152.31 and 175.79 pounds.

b) 154.54 to 173.56. We can be 90% confident that the mean weight of the population of all males 18–24 years old who are 70 inches tall is somewhere between 154.54 and 173.56 pounds.

PROBLEM 15.6

a) From the output shown in the answer to Problem 15.5, we see that a 95% prediction interval is from 124.96 to 203.14. We can be 95% certain that the weight of a randomly selected male 18–24 years old who is 70 inches tall will be somewhere between 124.96 and 203.14 pounds.

b) 132.38 to 195.72. We can be 90% certain that the weight of a randomly selected male 18–24 years old who is 70 inches tall will be somewhere between 132.38 and 195.72 pounds.

PROBLEM 15.7

STEP 1 The null and alternative hypotheses are

H_0: $\rho = 0$ (height and weight are linearly uncorrelated)
H_a: $\rho > 0$ (height and weight are positively linearly correlated)

STEP 2 $\alpha = 0.05$
STEP 3 We have $n = 11$, so that df $= 9$. Also $\alpha = 0.05$. Using Minitab's INVCDF command, we found that $t_{0.05} = 1.8331$. This is the critical value for a right-tailed test.
STEP 4 In Problem 14.5, we used Minitab's CORRELATION command to obtain the linear correlation coefficient, $r = 0.662$. Therefore, the value of the test statistic is

$$t = \frac{r}{\sqrt{\dfrac{1-r^2}{n-2}}} = \frac{0.662}{\sqrt{\dfrac{1-(0.662)^2}{11-2}}} = 2.6498$$

STEP 5 Since this is greater than the critical value of 1.8331, we reject H_0. Evidently, the variables, height and weight are positively correlated. Also from the printout for Problem 14.5 on page A-26, the P-value is listed as 0.026, confirming the rejection of H_0.

PROBLEM 15.8

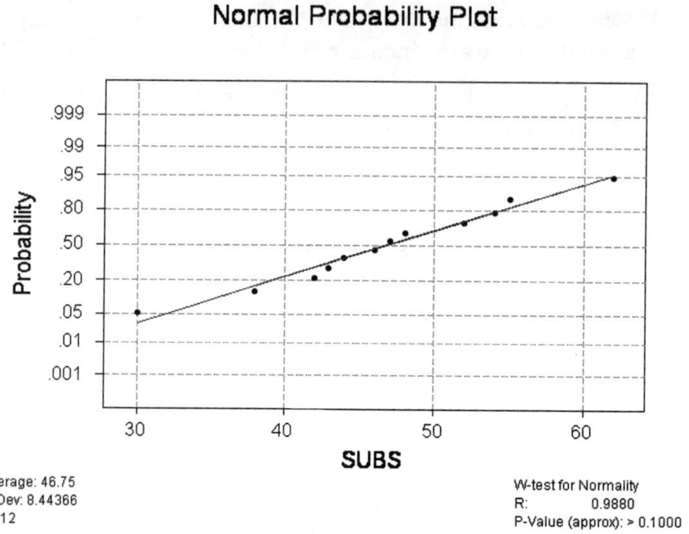

From the figure, you see that the correlation between NSCORES and AGI is 0.988. This produces a P-value of 0.10 and so you cannot reject the null hypothesis. That is, the data do not provide sufficient evidence to conclude that subscriptions are not normally distributed. This is consistent with the linear pattern of the foregoing plot.

Lesson 16

PROBLEM 16.1 See the solution to Problem 11.5 on page A-22

PROBLEM 16.2 Minitab output:

```
Analysis of Variance
Source      DF        SS        MS        F        P
Factor       2    115.11     57.56     8.53    0.003
Error       15    101.17      6.74
Total       17    216.28
                                    Individual 95% CIs For Mean
                                    Based on Pooled StDev
Level        N      Mean     StDev  ---+---------+---------+---------+---
POLICY1      6    22.500     2.168  (--------*-------)
POLICY2      6    23.167     2.483     (------*--------)
POLICY3      6    28.167     3.061                (--------*------)
                                    ---+---------+---------+---------+---
Pooled StDev =     2.597            21.0      24.0      27.0      30.0
```

With a P-value of 0.003, reject H_0. The data provide sufficient evidence to conclude that mean sales resulting from the different policies are not all the same. From the printout, $SST = 216.28$, $SSTR = 115.11$, and $SSE = 101.17$.

PROBLEM 16.3

Minitab output:

```
Tukey's pairwise comparisons

    Family error rate = 0.0500
Individual error rate = 0.0203

Critical value = 3.67

Intervals for (column level mean) - (row level mean)

           POLICY1     POLICY2

 POLICY2    -4.558
             3.224

 POLICY3    -9.558      -8.891
            -1.776      -1.109
```

From the confidence interval endpoints, we may conclude that there is no significant difference between mean sales for Policy 1 and Policy 2; it appears that mean sales for Policy 3 are less than those for Policy 1 and Policy 2.

PROBLEM 16.4

Minitab output:

```
Kruskal-Wallis Test on SALES

POLICY      N    Median    Ave Rank        Z
POLICY1     6     22.00         6.4    -1.73
POLICY2     6     23.50         7.5    -1.12
POLICY3     6     28.00        14.6     2.86
Overall    18                   9.5

H = 8.28  DF = 2  P = 0.016
H = 8.39  DF = 2  P = 0.015 (adjusted for ties)
```

Since the P-value is 0.016, the data provide sufficient evidence to conclude that mean sales resulting from the different policies are not all the same.